Uncertainty and Artificial Intelligence

Artificial Intelligence in Mechanics Set

coordinated by
Abdelkhalak El Hami

Volume 1

Uncertainty and Artificial Intelligence

Additive Manufacturing, Vibratory Control, Agro-composite, Mechatronics

Edited by

Abdelkhalak El Hami

WILEY

First published 2023 in Great Britain and the United States by ISTE Ltd and John Wiley & Sons, Inc.

ISTE Ltd
27-37 St George's Road
London SW19 4EU
UK

www.iste.co.uk

John Wiley & Sons, Inc.
111 River Street
Hoboken, NJ 07030
USA

www.wiley.com

Any opinions, findings, and conclusions or recommendations expressed in this material are those of the author(s), contributor(s) or editor(s) and do not necessarily reflect the views of ISTE Group.

Library of Congress Control Number: 2023944019

British Library Cataloguing-in-Publication Data
A CIP record for this book is available from the British Library
ISBN 978-1-78630-942-6

Contents

Chapter 6. Study of the Influence of Noise and Speed on the Robustness of Independent Component Analysis in the Presence of Uncertainty

Dorra BEN HASSEN, Anoire BEN JDIDIA, Mohamed Slim ABBES, Fakher CHAARI and Mohamed HADDAR

Chapter 7. Multi-Objective Optimization Applied to a High Electron Mobility Transistor

Rabii EL MAANI, Abdelhamid AMAR, Bouchaïb RADI and Abdelkhalak EL HAMI

Preface

Today, computing, along with artificial intelligence (AI), is moving towards complete communication between all computerized systems. AI is a representation of human intelligence that depends on the creation and application of algorithms in a precise computerized environment. Its objective is to allow computers to act like human beings. This type of technology needs computer systems, data with management systems and advanced algorithms capable of being used by AI.

In mechanics, AI offers several possibilities for mechanical construction, predictive maintenance, monitoring of installations, and in the fields of robotics, additive manufacturing, vibratory control and agro-composites.

AI needs a certain amount of data and a high processing capacity. It plays a role in a great number of fields, more specifically in enormous companies capable of using such technology. Its principal purpose is to improve the working conditions of employees in order to diminish risk.

The objective of this book is to explore the uncertainty of AI in mechanical problems. It is made up of seven chapters.

Chapter 1 applies one of the methods of AI called independent component analysis (ICA) to evaluate the power used during a circular operation so that we can estimate the energy used by a milling machine. This method is a technique of blind division of sources and is based on inverse problems. Its robustness has been demonstrated in several fields, for example, its use in approximating road surfaces using the knowledge of vehicular vibration or in the identification of defects related to gear systems.

The only thing required is knowledge concerning the dynamic response, for example, accelerations in response estimates. This chapter presents two digital models meant for estimating the power used by the spindle and the table of a mill during a circular milling operation. The method relies only on receivers that capture the vibratory responses such as movement or perhaps accelerations, and then estimates the additional effort made by the machine tool.

Chapter 2 uses several fields to define the best strategies for resolving a specific problem in one of the many applications of AI. The application is related to the use of AI during maintenance in additive manufacturing. However, the specific problem is associated with the existence of uncertainty in the performance of AI for this type of application. Several components are examined: uncertainty, AI, maintenance and additive manufacturing. The concept of uncertainty is first addressed separately in order to provide the reader with a clear explanation of this component. The study is made up of two threads: the first thread represents a proposed strategy, while the second thread is related to a specific application.

The strategy can be summarized as taking uncertainty into account when AI makes decisions. The degree of decisions made by AI can affect the result of the application either directly or indirectly. There are certain intervals for making the right decisions, which can be taught to the computer that is being used to avoid making the wrong decisions.

The principal objective of this chapter is to address questions of uncertainty in order to contribute to the industrialization of technology in additive manufacturing. The industrialization of additive manufacturing must implement several studies to prepare for different kinds of errors and the concept of uncertainty. A high rate of failure raises the total cost, which can be a major obstacle to the industrialization of technology in additive manufacturing.

Chapter 3 looks into the question of the durability of bio-sourced materials for industrial applications. The requirements and expectations of clients are becoming greater and greater with respect to the use of agro-composite components, which undergo mechanical uses and variations in temperature and humidity, repeated throughout their use, leading to a degradation of the mechanical properties and an accelerated aging of the vegetable fibers. The development of agro-composites for industrial applications requires precise information concerning the processes causing the damage in order to better predict their lifespan. Indeed, when there is a process that causes damage to these materials, a transitory wave, resulting from the release of the energy that was stored in it, propagates from the source of the damage towards the surface of the material. This wave can be recorded by receptors attached to the material. In such a context, this chapter looks for ways to make use of artificially intelligent methods to analyze the acoustic emission data and predict the appearance of critical mechanisms in order to identify scenarios where damage is caused. The purpose of our work is to implement methods that essentially allow us to analyze AI data. It depends on the use of recent classification methods (either unsupervised as in k-averages or supervised as with neural networks). The purpose of this multi-parameter statistical analysis is to identify the meaning of the data obtained during the monitoring of acoustic emissions for damage to the polymer matrix composite materials being mechanically worn along one axis.

Chapter 4 presents an intelligence method based on model-free control for the ease of managing an entire active suspension system in a vehicle. In the proposed controller, algebraic estimators are created for approximating the total perturbance in three senses (vertical, oscillation, slope), including the external effect of the road, the nonlinearities of the system and the modeling error, if it exists. A direct method does not require an overall mathematical model with an easy implementation. The intelligence of the controller boils down to its capacity to react quickly and online in fractions of a second to reject all unnecessary types of perturbance without needing to know their model or their frequential characteristics.

Chapter 5 aims to propose a methodology for studying the reliability of mechatronic systems through the reliability of high electron mobility transistors (HEMT) by using AI in the thermal model and the reliability and statistical model. Thermal modeling using the finite element method is presented to observe the thermal behavior of the transistor, and the influence of the working environment, such as the dissipated power and reference temperature on the working temperature. It then develops a thermo-reliability coupling by integrating two models: thermal and statistical, the thermal model being developed with Comsol Multiphysics and the statistical model (reliability) with Matlab. This coupling allows us to estimate the thermal reliability of the HEMT and identify the influence of the working conditions on its reliability and performance.

Chapter 6 is devoted to studying how robust the intelligence method of independent component analysis (ICA) is when estimating a road surface for a quarter vehicle. To do this, the Monte Carlo stochastic technique is used in the presence of the inevitable variables of uncertainty: the mass of the vehicle, the stiffness of the spring and shock absorption. The effect of noise generated by wind is also considered. The convergence of the ICA intelligence method was evaluated when comparing real surfaces as defined by the ISO norm to the surfaces approximated with uncertainties. The results obtained prove the robustness of the ICA in the estimation of different road surfaces.

Chapter 7 focuses on the study of the high electron mobility transistor (HEMT), which is one of the most important components in high-power mechatronic systems. HEMT is the most commonly used technology in complex systems in general and in mechatronic systems in particular. Consequently, the optimization of this technology is a major stake for engineers and researchers in this field. In this chapter, we present a method for multi-objective optimizing applied to HEMT in order to improve its thermal and mechanical performance. The optimization process is based on the coupling of two methods: the finite element model using Comsol Multiphysics software and the coded optimization model on Matlab software. The second model is used to solve the problem of optimization by coupling it with the first model. The optimal values of the conception variable obtained from the application of the AI methods and from the optimization process allow us to optimize the thermomechanical behavior of the HEMT structure, making it more reliable.

Abdelkhalak EL HAMI
Rouen
July 2023

New Intelligence Method for Machine Tool Energy Consumption Estimation

1.1. Introduction

Modern life has led us to consume high levels of electric energy, which is required for our needs. Research has demonstrated that the role of industry in worldwide energy consumption will rise from 3,000 Mtoe in 2010 to 5,000 Mtoe in 2050. This increase is associated with high emissions from the greenhouse effect, which has a negative impact on the environment. This is why eco-production has become increasingly necessary: to limit the high consumption of energy by industry. To do this, researchers have used tools for estimating energy consumed by machine tools. This chapter applies an artificial intelligence method, called independent component analysis (ICA), to evaluate the power consumed during a peripheral milling operation.

Our life is dominated by new technologies that have led to a heightened consumption of electrical energy. For example, in the United States, 31% of all energy is consumed by industry. 90% of this consumption goes to manufacturing, of which 70% is used by machine tools (Zhou et al. 2016). From there, atmospheric CO_2 emissions are caused by the energy consumption of the manufacturing industry. According to Herzog (2009), 99% of environmental problems are caused by the consumption of energy in manufacturing processes. The harmful effects are enormous, and the

Chapter written by Dorra Ben Hassen, Anoire Ben Jdidia, Mohamed Taoufik Khbou, Mohamed Slim Abbes and Mohamed Haddar.

research by Herzog (2009) has indeed shown that manufacturing operations are responsible for 19% of these emissions. This is why the optimization of energy consumption by machine tools is urgent for protecting the environment. In order to minimize this consumption, various studies have relied on modeling the energy consumed by machine tools, especially during cutting operations. Let us cite the model created by Rajemi et al. (2010) as an example, which allows for the quantification of the required energy by turning operations. This model accounts for the lifetime of a tool. We also note the model proposed by Avram and Xirouchakis (2011) for quantifying the power consumed by the spindle and the table of a mill while cutting. They calculate the energy of the spindle by multiplying the force of the cut by the speed of the cut. The energy of the table, meanwhile, is calculated using the product of the cutting torque and the angular speed. The Kara and Li (2011) model was based on the material removal rate (MRR). Another model was established by Calvanese et al. (2013), which presented the energy of an axis feed. It is equal to the product of the effort made in cutting and the cutting speed. To obtain this effort, the average value of the width of the shaving must be determined. Taking this model as a basis, an improvement was presented by Alberteli et al. (2016). They make use of the constant angular position of the cutting tool. Other recent work has been published in the literature by Ben Jdidia et al. (2019b). This work aims to estimate the dynamic cutting energy consumed by the spindle and the table of a mill for resurfacing operations. To estimate the effort involved in cutting, the authors apply the method of finite elements, which is a complex method.

Therefore, this chapter aims to estimate the energy consumed by a mill by applying a new method, namely ICA. This method is a major technique for blindly separating sources. It is based on inverse problems and is a simple method whose robustness has been demonstrated in numerous fields. For example, Ben Hassen et al. (2019) and Chaabane et al. (2019) use this method in estimating road surfaces, using knowledge related to vehicular vibration. Taktak et al. (2012) demonstrate the efficiency of this method in identifying the defects associated with a gear system. It only requires the knowledge of the dynamic response, such as accelerations, for estimating the responses. This is the inverse method.

1.2. Mathematical model for estimating power consumption by the spindle and table of a mill

Figure 1.1 describes the machine tool studied (a mill) during a peripheral cutting operation. Two principal efforts are studied, the tangential component of the cutting effort $F_t(t)$ applied to the spindle and the effort made in advancing $F_x(t)$, applied to the table. These two efforts are considered dynamic. Indeed, one tooth, denoted i, that removes material has a momentary angular position called $\Phi i(t)$. A shaving whose width varies over time is generated. This width is formed by two components: one static, caused by the rigid body, and one dynamic, caused by the motion of the tool at time t and t-τ:

$$\phi_i(t) = \Omega t + \phi_p - \psi \qquad [1.1]$$

where Ω is the speed of the spindle rotation in-rpm, ϕ_p is the angle between two successive teeth and ψ is the cutting angle.

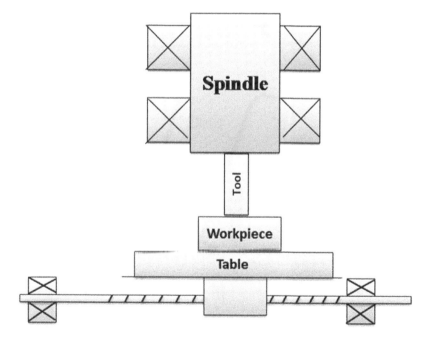

Figure 1.1. *Cutting system of a machine tool: a mill*

The power consumed by the spindle and the table during the machining operation is given by the following two equations:

$$P_{Table}(t) = F_x(t) \times V_f \qquad [1.2]$$

$$P_{Spindle}(t) = \frac{F_t(t) \times V_c}{60} \qquad [1.3]$$

where:

– V_c represents the cutting speed in m/min;

– V_f represents the feed speed in mm/min.

Figure 1.2 shows the distribution of cutting efforts during a circular machining operation (Romdhane 2017).

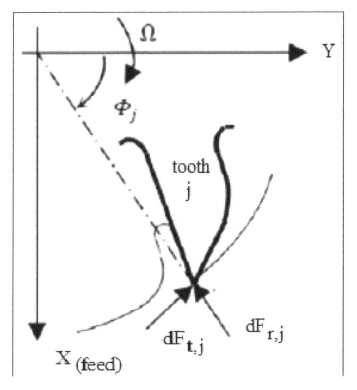

Figure 1.2. *Distribution of the cutting effort for a circular machining operation*

The projection of the differential cutting effort gives the following results:

$$dF_{x,i}(t) = -\cos(\phi_i(t))\,dF_{t,i}(t) - \sin(\phi_i(t))dF_{r,i}(t) \qquad [1.4a]$$

$$dF_{r,i}(t) = k_r dF_{t,i}(t) \qquad [1.4b]$$

$$dF_{t,i}(t) = k_t a_p h(t)dZ \qquad [1.5]$$

where $dF_{x,i}$, $dF_{r,i}$ and $dF_{t,i}$ are, respectively, the efforts of advancement, radial and tangential differentials, k_r is the specific radial pressure of cutting, k_t is the specific tangential pressure of cutting, a_p is the depth of the cut in mm and $h(t)$ designates the variable width of the shavings as a function of time, which is determined using the following equation:

$$h(t) = \big(u_x(t) - u_x(t - \tau)\big)\sin(\phi_i(t)) - \big(u_y(t) - u_y(t - \tau)\big)\cos(\phi_i(t)) \qquad [1.6]$$

We note that:

− $u_x(t)$ and $u_y(t)$ represent the displacement of the tool at time t;

− $u_x(t-\tau)$ and $u_y(t-\tau)$ represent the displacement at time t-τ.

The total cutting effort exerted by all of the teeth over the machined object can be written as follows:

$$F_x(t) = \sum_{i=1}^{N}\sum_{k=1}^{N_f} dF_{x,i}(t) \qquad [1.7]$$

$$F_t(t) = \sum_{i=1}^{N}\sum_{k=1}^{N_f} dF_{t,i}(t) \qquad [1.8]$$

$$F_r(t) = \sum_{i=1}^{N}\sum_{k=1}^{N_f} dF_{r,i}(t) \qquad [1.9]$$

where:

− N represents the number of teeth;

− N_f is the number of cutting elements.

It is very clear from the previous equations that the determination of the cutting effort is essential for quantifying the two types of power. To do this, the artificial intelligence method ICA is applied to estimate the variable tangential and cutting efforts.

1.3. ICA method

ICA is an important method for the blind separation of sources. It makes it possible to decompose a random signal X(t), in this case the motion of the nodes, into a combination of statistically independent components that are the estimated cutting efforts (Abbes et al. 2011). The vector of observed signals can be written as (Hassen et al. 2019):

$$X(t) = \{A\}\{S(t)\} \qquad [1.10]$$

where A is the mixing matrix and S represents the source signals.

The stages of application in ICA are provided in detail below.

1.3.1. *Pretreatment of signals*

The vector X(t) of observed signals undergoes pretreatments, namely centering and whitening, in order to make ICA evaluation easier.

1.3.1.1. *Centering*

Centering consists of subtracting from vector X its average vector $m = E\{X\}$ so that its average becomes zero and then S can also have an average of zero.

1.3.1.2. *Whitening*

The second pretreatment in ICA is the whitening of observed variables. This technique makes it possible to eliminate the noise of a signal. This vector has a matrix with unit covariance. We then get:

$$E\{XX^*\} = I \qquad [1.11]$$

So, to whiten a signal X, we calculate its covariance matrix as follows:

$$R_X = E\left\{X(t)X^*(t)\right\} = E\left\{AS(t)(AS(t))^*\right\}$$
$$= AA^* \underbrace{E\left\{S(t)S^*(t)\right\}}_{I} = AA^* = UDU^T \qquad [1.12]$$

where:

– U is the orthogonal matrix of the eigenvectors of matrix R_X;

– D is the diagonal matrix of its eigenvectors of matrix R_X.

We thus get:

$$R_X = E\left\{WX(t)(WX(t))^*\right\} = WW^* \underbrace{E\left\{X(t)X^*(t)\right\}}_{UDU^T} = I \qquad [1.13]$$

We then determine the whitening matrix as follows:

$$W = D^{-\frac{1}{2}}U^T \qquad [1.14]$$

1.3.2. Separation

After undergoing the pretreatment, the vector for observed signals X(t) is used as a point of entry to ICA in order to separate the sources using the following steps.

1.3.2.1. Maximization of the kurtosis function

Kurtosis is a tool that finds non-Gaussian components and their place in the field of frequency. It is defined as the normalized marginal fourth-order cumulant (Achard 2003; Zarzoso and Comon 2008):

$$K(w) = \frac{E\left\{|y|^4\right\} - 2E^2\left\{|y|^2\right\} - \left|E\left\{y^2\right\}\right|^2}{E^2\left\{|y|^2\right\}} \qquad [1.15]$$

We can see that this criterion is indifferent to the scale factor, in other words: $K(\lambda w)=K(w)$ for $\lambda \neq 0$.

Since the scale is generally unimportant, we can apply normalization. $\|W\|=1$.

To maximize the kurtosis, we must find the direction of the absolute contrast of kurtosis, defined in Zarzoso and Comon (2008) by the following equation:

$$\mu_{opt} = \arg\max_{\mu} \left| K(w + \mu g) \right| \qquad [1.16]$$

The sought-out direction is given by the gradient, $g = \nabla_w K(w)$, defined by:

$$\nabla_w K(w) = \frac{4}{E^2\{|y|^2\}}$$
$$\left\{ E\{|y|^2 y x\} - E\{yx\} E\{y^2\} - \frac{(E\{|y|^4\} - |E\{y^2\}|^2) - E\{yx\}}{E\{|y|^2\}} \right\} \qquad [1.17]$$

1.3.2.2. *Determination of the optimal step*

The optimal step, defined by equation [1.16], can be determined by calculating the roots of a fourth-degree polynomial. To find the optimal step, robust ICA goes through the following steps (Cao and Murata 1999):

1) Calculate the coefficients of the polynomial for the optimal step defined by:

$$p(\mu) = \sum_{k=0}^{4} a_k \mu^k \qquad [1.18]$$

Coefficients $\{a_k\}_{k=1}^{4}$ can be obtained at each iteration using observed signals and the actual values of w and g. Their expressions are found in Appendix A.4 of Zarzoso and Comon (2009).

2) Calculate the polynomial roots of the optimal step $\{u_k\}_{k=1}^4$.

3) Choose the root that leads to the absolute maximum contrast over the length of the direction g.

$$\mu_{opt} = \arg\max_k \left| k(w + \mu_k g) \right| \qquad [1.19]$$

4) Update w:

$$w^+ = w + \mu_{opt} g \qquad [1.20]$$

5) Normalize w+:

$$w^+ \leftarrow \frac{w^+}{\left\| w^+ \right\|} \qquad [1.21]$$

The algorithm can be stopped when

$$\left| 1 - \left| w^T w^+ \right| \right| \prec \varepsilon \qquad [1.22]$$

where ε is a small, static constant that can be defined by: $\varepsilon = \eta/T$, where $\eta < 1$.

1.3.2.3. Deflation

Deflation makes it possible to derive sources one after another. This approach is based on two stages. The first consists of identifying a source in the mix and the second subtracts the contribution of this source to the mix. The location of the updated derived vector is indeed constrained to the orthogonal sub-space of the previously derived vectors. Or rather, we can use the principle of linear, deflated regression. This consists of calculating the contribution of the source, estimated through observation using the minimum solution of the squared error average for the problem of linear regression $X = X - \hat{h}\hat{s}$, which is given by:

$$\hat{h} = \arg\max_h E\left\{ \left\| X - h\hat{s} \right\|^2 \right\} = \frac{E\{X\hat{s}\}}{E\{|\hat{s}|^2\}} \qquad [1.23]$$

The observations are thus recalculated as follows, and the approach will be repeated N times until all sources are extracted:

$$X \leftarrow X - \hat{h}\hat{s} \tag{1.24}$$

In brief, ICA is used in our case to estimate the effort in cutting as presented in the flowchart in Figure 1.3.

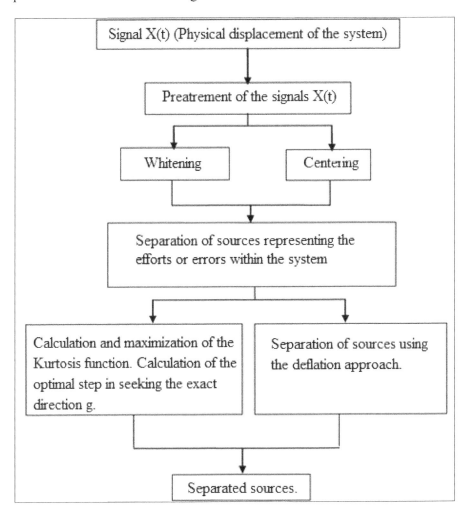

Figure 1.3. *The principal stages of the separation program*

1.4. Results and discussion

In this section, the application of the ICA method is presented in Figure 1.3. The vibratory responses of the cutting system are obtained by solving the motion equation [1.25] using the method of finite elements (Figure 1.4). Afterwards, they are inserted in the ICA as if they were the observed signals used to evaluate the tangential cutting effort $F_t(t)$ and the effort of moving through the cut $F_x(t)$. Finally, the power is deduced from the solution to the motion equation.

The method of finite elements was applied to the spindle (Figure 1.4) in order to develop the following motion equation:

$$[M]\{\ddot{Q}\} + 2\Omega[G]\{\dot{Q}\} + \left([K] - \Omega^2[C]\right)\{Q\} = \{F_c(t,Q)\} \qquad [1.25]$$

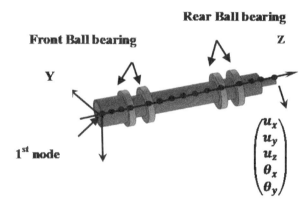

Figure 1.4. *Finite element model of the spindle. For a color version of this figure, see www.iste.co.uk/elhami/uncertainty.zip*

where:

– [M], [G], [K] and [C] are the mass, gyroscopic, rigidity and shock absorption matrices, respectively;

– vector Q presents the motion at each moment;

– Fc is the vector for variable cutting efforts.

The equation of motion is solved using the Newmark method coupled with the Newton–Raphson method in order to estimate the temporal tooltip displacements.

The equation is solved using a cutting speed equal to 140 m/min, a passage depth equal to 1 mm and a tooth progression equal to 0.1 mm/tooth.

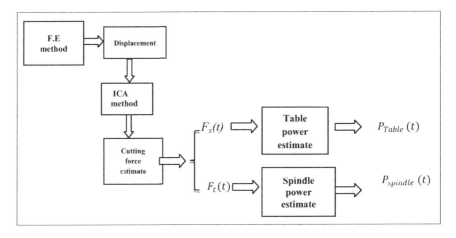

Figure 1.5. *Implementation of the ICA method*

A parametric study was performed to choose the best estimate of cutting efforts. The results showed that the closer one is to the ball-bearings, the more precise the estimate becomes. We note that in this study, bearings are introduced into our model as limits.

Figure 1.6 shows a comparison of the tangential and advancement efforts estimated using the ICA and real methods.

The results in Figure 1.6 demonstrate that the ICA allows us to estimate the cutting effort for the circular machining operation. There is strong agreement between the real and estimated efforts.

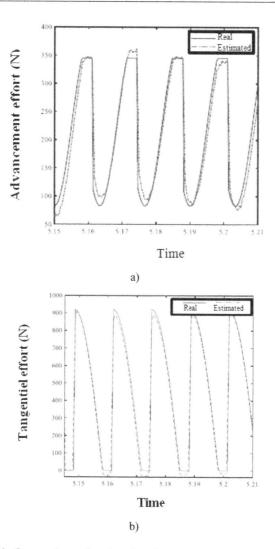

Figure 1.6. *Comparison of real and estimated tangential efforts (a) and real and estimated advancement efforts (b). For a color version of this figure, see www.iste.co.uk/elhami/uncertainty.zip*

For validating the choice of nodes, the MAC (Modal Assurance Criteria) calculations (equation [1.26]) and the relative error (equation [1.27]) between the real and estimated efforts are calculated:

$$MAC = \frac{\left(F_{real}^T F_{estimated}\right)^2}{\left(F_{real}{}^T F_{estimated}\right)\left(F_{estimated}{}^T F_{real}\right)} \qquad [1.26]$$

$$E_r(\%) = 100\frac{|F_{real}-F_{estimated}|}{F_{real}} \qquad [1.27]$$

Figure 1.7 illustrates the amplitude of MAC for each of the efforts, tangential and advancement.

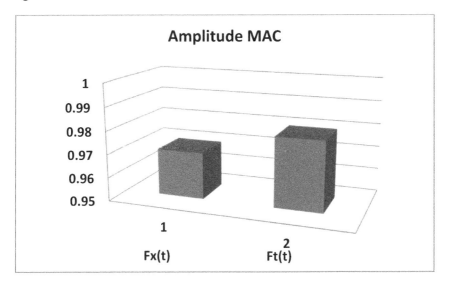

Figure 1.7. *Calculation of the MAC amplitude for tangential Ft(t) and advancement Fx(t) efforts. For a color version of this figure, see www.iste.co.uk/elhami/uncertainty.zip*

The MAC for each effort is close to 1, which proves the validity of the estimate obtained through the ICA method.

The calculation of the relative error gives the results summarized in Table 1.1.

	Fx(t)	Ft(t)
$E_r(\%)$	0.6	1.6

Table 1.1. *Relative errors*

It is very clear that the relative error values are infinitely small, which leads us to conclude that the estimate obtained by the ICA method can indeed be validated.

1.5. Conclusion

This chapter presents two numerical models meant for estimating the power consumed by the spindle and table of a mill during a circular machining operation. The advantage of the model developed is the consideration of the nonlinear behavior of the cutting efforts during material removal. The most important idea lies in the use of an intelligent ICA method for estimating the cutting efforts and later provides an estimate of the cutting power consumed by the machining system. The results obtained confirm that this method, which is based on artificial intelligence, can be considered a tool in the field of manufacturing for helping decide on the selection of cutting parameters. The method relies only on captors to identify the vibratory responses such as displacement or acceleration, and it then estimates the increased efforts of the machine tool.

1.6. References

Abbes, M.S., Chaabane, M.M., Akrout, A., Fakhfakh, T., Haddar, M. (2011). Vibratory behavior of a double panel system by the operational modal analysis. *International Journal of Modeling, Simulation, and Scientific Computing*, 2(4), 459–479.

Achard, S. (2003). Mesures de dépendance pour la séparation aveugle de sources. Application aux mélanges post non linéaires. PhD Thesis, Université Joseph-Fourier-Grenoble I, Saint-Martin-d'Hères.

Albertelli, P., Keshari, A., Matta, A. (2016). Energy oriented multi cutting parameter optimization in face milling. *Journal of Cleaner Production*, 137, 1602–1618.

Avram, O.I. (2010). Machine tool use phase: Modeling and analysis with environmental considerations. PhD Thesis, École Polytechnique Fédérale de Lausanne, Lausanne.

Avram, O.I. and Xirouchakis, P. (2011). Evaluating the use phase energy requirements of a machine tool system. *Journal of Cleaner Production*, 19(6–7), 699–711.

Ben Hassen, B., Miladi, M., Abbes, M.S., Baslamisli, S.C., Chaari, F., Haddar, M. (2019). Road profile estimation using the dynamic responses of the full vehicle model. *Applied Acoustics*, 147, 87–99.

Ben Jdidia, A., Bellacicco, A., Hentati, T., Barkallah, M., Khabou, M.T., Rivier, A., Haddar, M. (2019a). Modelling of axis feed consumed energy for sustainable manufacturing. *Journal of the Chinese Institute of Engineers*, 42(5), 377–384.

Ben Jdidia, A., Hentati, T., Bellacicco, A., Khabou, M.T., Riviere, A., Haddar, M. (2019b). Energy consumed by a bearing supported spindle in the presence of a dynamic cutting force and of defects. *Comptes rendus mécanique*, 347(10), 685–700.

Calvanese, M.L., Albertelli, P., Matta, A., Taisch, M. (2013). Analysis of energy consumption in CNC machining centers and determination of optimal cutting conditions. In *Re-engineering Manufacturing for Sustainability*, Nee, A.Y.C., Song, B., Ong, S.-K. (eds). Springer, Singapore.

Cao, J. and Murata, N. (1999). A stable and robust ICA algorithm based on t-distribution and generalized Gaussian distribution models. In *Neural Networks for Signal Processing IX: Proceedings of the 1999 IEEE Signal Processing Society Workshop (Cat. No. 98TH8468)*, IEEE, 283–292.

Chaabane, M.M., Ben Hassen, D., Abbes, M.S., Baslamisli, S.C., Chaari, F., Haddar, M. (2019). Road profile identification using estimation techniques: Comparison between independent component analysis and Kalman filter. *Journal of Theoretical and Applied Mechanics*, 57.

Hentati, T., Barkallah, M., Bouaziz, S., Haddar, M. (2016). Dynamic modeling of spindle-rolling bearings systems in peripheral milling operations. *Journal of Vibroengineering*, 18(3), 1444–1458.

Herzog, T. (2009). World greenhouse gas emissions in 2005. Working paper, World Resources Institute, Washington, D.C.

IEC (2007). Standard methods for determining losses and efficiency from tests (excluding machines for traction vehicles). International standard, International Electrotechnical Commission, Geneva.

Kara, S. and Li, W. (2011). Unit process energy consumption models for material removal processes. *CIRP Annals*, 60(1), 37–40.

Rajemi, M.F., Mativenga, P.T., Aramcharoen, A. (2010). Sustainable machining: Selection of optimum turning conditions based on minimum energy considerations. *Journal of Cleaner Production*, 18(10–11), 1059–1065.

Romdhane, T.F. (2017) *Systèmes intelligents et communicants – les réseaux de capteurs sans fil*. Noor Publishing, Riga.

Taktak, M., Tounsi, D., Akrout, A., Abbès, M.S., Haddar, M. (2012). One stage spur gear transmission crankcase diagnosis using the independent components method. *International Journal of Vehicle Noise and Vibration*, 8(4), 387–400.

Zarzoso, V. and Comon, P. (2008). Robust independent component analysis for blind source separation and extraction with application in electrocardiography. In *2008 30th Annual International Conference of the IEEE Engineering in Medicine and Biology Society*, IEEE, 3344–3347.

Zarzoso, V. and Comon, P. (2009). Robust independent component analysis by iterative maximization of the kurtosis contrast with algebraic optimal step size. *IEEE Transactions on Neural Networks*, 21(2), 248–261.

Zhou, L., Li, J., Li, F., Meng, Q., Li, J., Xu, X. (2016). Energy consumption model and energy efficiency of machine tools: A comprehensive literature review. *Journal of Cleaner Production*, 112, 3721–3734.

2

Uncertainty and Artificial Intelligence: Applications to Maintenance in Additive Manufacturing

2.1. Introduction

This chapter combines several fields to define the best strategies for resolving a specific problem in one of the numerous applications of artificial intelligence. The application is related to the use of artificial intelligence for maintenance in additive manufacturing.

The specific problem, however, is related to the existence of uncertainty in the performance of artificial intelligence for this type of application. In this chapter, we will discuss several aspects: uncertainty, artificial intelligence, maintenance and additive manufacturing. The concept of uncertainty is first discussed separately in order to provide the reader with a clear explanation of this component.

After this, we will discuss the aspects in relation to each other following their treatment in the literature.

Consequently, the study contains two threads: the first represents a proposed strategy, while the second thread is related to a specific application. The strategy can be described as a way of taking into account the uncertainty involved in decision-making by artificial intelligence.

Chapter written by Ghais KHARMANDA, Hicham BAAMMI and Abdelkhalak EL HAMI.

In fact, the degree of decision-making by artificial intelligence can affect the result of an application in a direct or indirect way. There are certain intervals for making good decisions, which can be taught to the machine in use in order to avoid the wrong decisions being made.

Given the uncertainty in conception and process, computers could receive the best possible instructions for completing their tasks, leading to an improvement in productivity. Moreover, the efficient and precise decisions made by artificial intelligence raise the sense of security in the process itself and improve the quality of the product.

By applying this strategy to one or several stages of additive manufacturing, it could increase the possibility of providing a clear means for industrializing this technology in the future.

The principal objective of this chapter is to respond to questions of uncertainty in order to contribute to the industrialization of additive manufacturing technology. The industrialization of additive manufacturing must involve research that confronts different failures that could arise as well as the concept of uncertainty.

A high failure rate raises the total cost, which can be a major obstacle to the industrialization of the additive manufacturing technology. Thus, the different possible failures must first be identified and then addressed. Moreover, uncertainty must be taken into account at several levels: conception, materials, additive manufacturing procedures (machine maintenance, manufacturing parameters, etc.).

2.2. Integration of uncertainty

2.2.1. *Definition of uncertainty*

Uncertainty can simply be understood as the impossibility of defining exact future results due to a limitation on existing knowledge. It can refer to failure for which information is barely available. This uncertainty can exist in the design and/or the process. Uncertainty is related to variability, which can come from various risks of two principal origins: 1) intrinsic risk (operational constraints), which can affect the quality of the product, is principally related to operational constraints (internal forces, movements, etc.) and quality constraints (material properties, surface qualities, measurement precision, etc.); and 2) extrinsic risks (quality constraints), which are principally related to the environment and can have two points of origin: climactic stress (temperature, humidity, pressure, etc.) and mechanical stress (charge, forced deformation, etc.) (Tebbi et al. 2003).

2.2.2. Types of uncertainty

According to Bradley and Drechsler (2014), there are three types of uncertainty: 1) ethical uncertainty; 2) optional uncertainty; and 3) state-space uncertainty. An ethical uncertainty appears if the agent cannot attribute precise uses to the consequences. The optional uncertainty appears when the agent does not know which consequence corresponds to an act at each state. Finally, the state-space uncertainty appears when the agent does not know how to construct an exhaustive space. These types of uncertainty are characterized by taking three dimensions into account: the nature, the object and the severity.

2.3. Uncertainty in artificial intelligence

2.3.1. Concepts and challenges in artificial intelligence

The objective of artificial intelligence is to use computers to complete dangerous or fastidious tasks while taking different principles of human intelligence into account. Here, three principal domains should be formalized: calculations, probabilities and logic. Calculations are used to analyze the problems which can be calculated using the theory of complexity. Then, the concept of probability seeks to manage uncertainty in this domain. Finally, to make a decision, a logical theory is needed to combine the theories of probability and utility. It is difficult, however, to express everything using the theory of logic, which leads to a consideration of the concept of uncertainty in artificial intelligence. Moreover, one of the essential applications of artificial intelligence is additive manufacturing, in which uncertainty must be taken into account at several levels. An efficient maintenance strategy can also be used to improve the productivity of additive manufacturing.

Artificial intelligence is related to data science, whose objective is to extract useful information from a set of data. Thus, work in artificial intelligence and data science is continuously being developed and requires more talent to respond to the needs of industry. Although technology has rapidly accelerated in industry, the integration of this evolving technology demands that universities align educational experience with evolving personal needs. There is an increasing and urgent recognition of the potential for racial prejudice and other forms of prejudice that can shape artificial intelligence and data science with great social and economic impacts on individuals and communities. The development of mutually beneficial collaborations between industry and academia could give rise to new ideas that simultaneously improve research and educational experiences while favoring diversity, equity and inclusion. To have a veritable social impact, intersectional approaches must also create trust in order to catalyze the progress of artificial intelligence and data science for the benefit of society. These high-impact interactions could lead to new courses, programs and research projects that would in turn improve the technologies of artificial intelligence/data science as well as the personnel (Washington 2022). Faced with uncertainty in artificial science, the objective is to make a decision while keeping time and costs in mind (reasonable times and costs).

Many decision-makers consider good decisions to be essentially based on data, facts and past experiences. The data of past experiences, however, is sometimes the synonym of a lack of improvement. These past experiences could be sub-optimal, and they should always be discussed. So, instructions or stages defined by experience can be optimized one by one, especially when it has to do with risk. Consequently, this observation is not considered, as it could lead to excessive confidence and reduce the possibility of involving risk management. Risk management is another element that can contribute to the success of this approach. Risks can also involve a degree of uncertainty and control the result and the consequences entirely. Nevertheless, in this chapter, we will only discuss decision-making in the context of uncertainty.

2.3.2. *Implementing artificial intelligence*

A simplified diagram of the implementation of artificial intelligence is presented in Figure 2.1. There are three generic phases: pre-implementation, implementation and post-implementation. The first phase is called "pre-implementation". It corresponds to the period during which an organization becomes aware of the existence of a technology. The second phase is called the "implementation", during which the members of the organization adopt an attitude with regard to the technology and begin activities which, in the end, lead them to choose whether to accept or reject the technology. The third phase, called "post-implementation", consists, for the organization, of seeking support for the decision made. The implementation of artificial intelligence is related to three factors: the people, the process and the technology (for more details, an interested reader should refer to Drmac (2022)).

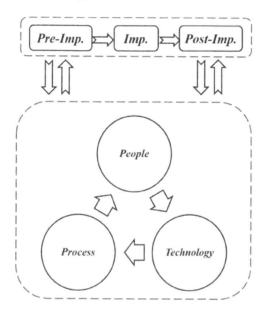

Figure 2.1. *Implementation of the artificial intelligence process. For a color version of this figure, see www.iste.co.uk/elhami/uncertainty.zip*

2.3.3. *Decision-making in the case of uncertainty*

As mentioned in the introduction, there are three principal areas that must be formalized in artificial intelligence: calculations, probabilities and logic. With respect to the logic stage, it is difficult to express everything using the theory of logic, which leads us to consider the concept of uncertainty in artificial intelligence. Making the right decisions quickly is a great challenge, even in the best of conditions (Janse 2021).

We know that by collecting enough data, selecting (or developing) the appropriate formula and using previous experience (past mistakes), the correct responses are arrived at automatically. However, the process functions differently, and uncertainty must really be taken into account and estimated.

2.4. Artificial intelligence in additive manufacturing

2.4.1. *Notions and issues in additive manufacturing*

Additive manufacturing, also known as 3D printing or rapid prototyping, is a technology that is taking off, which provides the possibility of transforming manufacturing for the next generation. 3D printing constructs solid objects layer by layer in the same way a 2D printer does, with the "printed" layers getting laid one over the other (Gebhardt and Fateri 2013). There are several techniques for additive manufacturing, such as stereolithography (SL), fused deposition modeling (FDM), direct metal laser sintering (DMLS) and electron beam melting (EBM). The procedures of additive manufacturing are based on the deposit or the solidification of materials layer by layer, which reduces the limitations of geometric complexity to a large extent (Liu et al. 2018). The technology of additive manufacturing has elicited a great deal of interest in the academic sphere as well as in industry due to its capacity for creating complex forms with specifiable material properties (Gao et al. 2015). It can also be called a free-form manufacturing technique in which the step of optimizing the topology can be completed without considering the optimization process of the form or size (Kharmanda and Antypas 2020). Moreover, topology optimization can be used to resolve several problems in additive manufacturing (Pradel et al. 2018; Fu 2020; Kharmanda and Mulki 2022). Since 2010, several articles have summarized the use of optimization methods in the additive manufacturing process (Wong and Hernandez 2012; Frazier 2014; Shashi et al. 2017; Wiberg et al. 2019; Alfaify et al. 2020).

For complex geometric forms, several problems can arise during the use of conventional manufacturing. Additive manufacturing offers the possibility of executing the manufacturing process in a simple and efficient way despite several challenges, in particular high costs and the waste of materials. It is therefore necessary to rise to several challenges when integrating additive manufacturing into industrial production. Certain additive manufacturing companies point out, honestly, that even though almost three decades have passed since the additive manufacturing revolution, it has unfortunately not yet been industrialized. The principal problem can be presented as an attempt to provide good quality products. In general, a manufacturing strategy can be industrialized when many parts (thousands) can be manufactured. The rise in the failure rate in additive manufacturing brings about a rise in the cost of production. Consequently, it is preferable to simulate and/or to control the additive manufacturing process before reducing the failure probability. It is also possible to encounter unexpected problems during the additive manufacturing process. It is very difficult to ensure the properties of the final products, such as the roughness of the surface, the porosity, the fatigue, the durability, etc. However, we can introduce hybrid manufacturing here. Additive manufacturing makes it possible to obtain specific surface finishes, but this takes a great amount of time if the final surface is smooth. It is therefore preferable to make use of the conventional method since rapid techniques for ensuring the established parameters for surface roughness already exist. Surface finishing is important for preventing corrosion and Taylor displacement.

It is therefore currently difficult to industrialize the technology of additive manufacturing. Several problems must be resolved, one of which concerns the way in which we account for uncertainty in the industrialization of additive manufacturing.

2.4.2. Uncertainty in additive manufacturing

There are several sources of uncertainty when executing the additive manufacturing process, and it is important to evaluate the different risks resulting in uncertainty. For example, uncertainty in geometric models can lead to a manufacturing imperfection. Thus, certain deviations with respect to the geometry as it was imagined can arise (Liu et al. 2018). Another type of uncertainty corresponds to material uncertainties. Indeed, the properties of materials are associated with several parameters such as the construction direction, the orientation, the extrusion, etc. Uncertainty over the properties of the materials also affects the cost of manufacturing. It is therefore necessary to find a framework of efficient additive manufacturing to resolve this type of problem (Li and Tsavdaridis 2021; Ribeiro et al. 2021). There are generally several types of uncertainty to analyze. In our previous study, thermal uncertainty was studied in the context of preheating in the additive manufacturing process (Kharmanda 2022).

2.4.3. *Probable failure scenarios in additive manufacturing*

Uncertainty is related to failure scenarios, which must be first identified, and then the cause of these failure scenarios must be found. An indicator can be added to ensure that a certain value belongs to an interval of a given set of parameters. If the value of this parameter exceeds the interval, there is a failure. If the value goes beyond a given interval, the logic suggests that there could be several causes responsible for the deviation. This component probability needs to be replaced. In this section, we present some causes of such failure scenarios.

– The complexity of the geometry: support structures must be provided for certain complex geometries (overhanging features) during the additive manufacturing process. To reduce the waste of materials and repeated trials, it is recommended that a raft should be added, and the temperature of the platform should be raised during the manufacturing process.

– Quality of materials in additive manufacturing: to reduce the waste of materials and repeated trials, it is preferable to select a material of good quality.

– Preheating the extruder and the platform: certain manufacturers suggest using glue to improve the adhesion level. We consider that this solution could affect the quality of the platform in future operations. It is therefore preferable to increase the preheating temperature of the extruder and the platform (Kharmanda 2022).

– Heating the platform during the additive manufacturing process: to reduce the waste of materials and repeated trials, it is necessary to raise the temperature of the platform during the additive manufacturing process.

– Heating the extruder during the additive manufacturing process: to reduce the waste of materials and repeated trials, it is necessary to raise the temperature of the extruder during the process of additive manufacturing. There is, however, a limitation on this increase to reduce problems with the environment, energy, costs and product quality.

– Speed of filament: this parameter affects the process of additive manufacturing and the quality of the product. An appropriate speed of filament use makes it possible to obtain smooth surfaces and stability in the process.

– Homogeneity of dimensions: for large dimensions, there can be adhesion problems at certain points since the forces applied can produce a bending moment that leads to the separation of the raft from the platform.

2.4.4. Integration of artificial intelligence

Industrial applications of artificial intelligence can be classified into three main categories: the first is called "design and manufacturing", in which artificial intelligence is applied to the design optimization, the design acceleration and to additive manufacturing. The second group is called "customer operations", in which artificial intelligence is applied to damage prognostics, inspection services, network optimization, etc. The third group is called "service productivity automation", in which artificial intelligence is applied to dynamic optimization, IT automation, shop automation, verification and validation, etc. In this chapter, we concentrate on the first group, in which artificial intelligence is applied to additive manufacturing. Here, the objective is to use artificial intelligence to identify and exploit the relationships between input parameters and the manufacturing outcomes in the additive manufacturing process. According to PwC research and analysis, automated machine translation is expected from 2025 and creative art engines, automated 3D bioprinters, artificial wildlife habitats from 2030 (Washington 2022).

2.5. Predictive maintenance in additive manufacturing

2.5.1. Tendencies in maintenance technology

During the last seven decades, several waves in maintenance have appeared. These waves are divided into three categories: preventive (or scheduled), productive and predictive maintenance technologies (Poór et al. 2019). Each phase coincides with certain industrial developments, such as electric energy, automation, Internet, etc. The first tendency was called "preventive maintenance technology", which began in the 1950s with Japanese engineers; several actors formed a team that defined different procedures in order to prevent damage to equipment. It was a costly process, however, since numerous parts were replaced to delay a breakdown. This process caused unnecessarily long workdays. Thus, the preventive maintenance process improved the reliability of equipment before a breakdown could even happen (Gross 2006). The corresponding industrial period coincided with different developments in the following areas: mass production, assembly chains and electrical energy.

The second wave is called "productive maintenance technology", established in the 1960s and considered more professional than the first (Nakajima 1988). This technology required thorough knowledge of the reliability of the different components of the machine. The objective was to minimize accidents by improving the work environment, by cleaning machines and verifying the state they were in (zero downtime, zero errors and zero disturbances). The corresponding industrial period coincides with different developments in the following fields: automation, computers and electronics.

The third wave is called "predictive maintenance technology", which began in the 1990s (Butler 1996). The objective of this type of technology is to simulate how the product works and predict the different failure scenarios in order to prevent them from arising. Instead of using experts, a company must employ reliability and data engineers. The corresponding industrial period coincides with the different developments of the following fields: Cyber Physical Systems, the IoT (Internet of Things), networks, clouds and BDA (Big Data Analysis).

To compare these three waves, several characteristics can be found in the literature. There is a factor called the OEE (Overall Equipment Effectiveness) for measuring the effectiveness rate (Poór et al. 2019). Its value for preventive maintenance belongs in the range of 50–75%, and the range for productive maintenance is 75–90%. It is over 90%, however, for predictive maintenance. Another factor is the cost of maintenance. Among these three types, preventive maintenance is considered the most expensive and requires a high level of work. Moreover, it is characterized by excessive usage, which can affect the performance of the system. Productive maintenance is more advanced than preventive maintenance, while the most advanced type is predictive maintenance, which lets us improve certain criteria such as quality, productivity, etc., while reducing maintenance costs.

2.5.2. Diagnostic and prognostic models

A complete diagram of the stages of diagnostic/prognostic is illustrated in Figure 2.2. The diagnostic stage includes three sub-stages – fault detection, fault isolation and fault identification – while the prognostic stage only contains two sub-stages – predicting the remaining useful time and estimating the confidence interval.

The fault detection and isolation stages are related to one or several components or subsystems, while the fault identification stage is related to the fault scenario. In the diagnostic stage, the stage where the remaining useful time is predicted consists of determining the time left before failure occurs, while estimating the confidence interval is related to the effect of different factors on the prediction of the useful time that remains (for more details, the reader should refer to Sikorska et al. (2011)). At the last stage, uncertainty analysis plays an important role in determining confidence intervals. Moreover, by increasing the number of factors affecting the prediction of useful time left, we can improve the performance of artificial intelligence in making the right decisions.

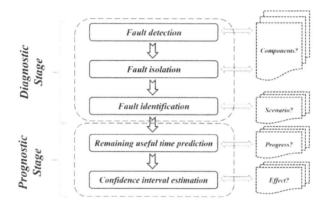

Figure 2.2. *Diagnostic and prognostic stages. For a color version of this figure, see www.iste.co.uk/elhami/uncertainty.zip*

2.5.3. *Implementation in additive manufacturing*

Predictive maintenance can be used to monitor the performance of parts manufactured in an additive way. It is possible to predict when the parts will degrade or fail and need to be replaced (Henry et al. 2020). Parts manufactured in an additive way display certain attributes due to the manufacturing process itself, which, if not controlled, can bring about the degradation of these parts in a way that is not seen for parts manufactured in the traditional way. In this way, it is generally understood that the use of these parts, in current cases, is meant to be a stopgap to fill a critical request. This can guarantee operational availability in increasingly contested environments where the immediate demand for replacement parts imposes constraints on the time required to deliver them from far away. With parts manufactured in an additive manner, the imperfections are generally attributed to empty spaces and geometric variation between the digital model and the printed item. The receiver offers a way of following individual parts, predicting the failure points and replacing them before they break down. By installing the replacement at lower lifetime, we highly reduce the risks, which leads to a high-trust situation. The objective is to understand the specific performance of a material printed in 3D by monitoring and detecting its performance and then its degradation.

2.6. Proposed strategy and applications

This section covers the matter of the specific application and the advanced system of quality monitoring being discussed.

2.6.1. *Examples of uncertainty in additive manufacturing*

It is very important to know that each situation contains certain levels of uncertainty. We present here two cases of uncertainty in additive manufacturing. In the first, the cause of failure is not clear, while in the second, it can be observed. Figure 2.3 shows an example of uncertainty in additive manufacturing during the production of a candy box. For the same starting parameters, we sometimes obtain a failure scenario that can happen between two successful trials. In this case, the cause of the failure is not clear. It could be due to a change in temperature (of the extruder and/or the platform), a change in filament speed, another non-uniformity in the filament (related to the quality of the raw material) and/or another cause.

Figure 2.4 shows another example of uncertainty in additive manufacturing during the production of a column base (same starting parameters).

Figure 2.3. *Successful trial of a candy box and unsuccessful trial (the product can be found on 3d-printing-4u.com). For a color version of this figure, see www.iste.co.uk/elhami/uncertainty.zip*

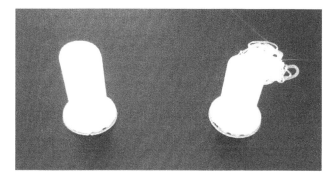

Figure 2.4. *Successful trial of a column base and unsuccessful trial (the product can be found on 3d-printing-4u.com). For a color version of this figure, see www.iste.co.uk/elhami/uncertainty.zip*

The cause of the failure is a perturbation in the distribution of the material at the lower layer of the raft (Figure 2.5), which caused instability during the manufacturing process. In this way, the moment increases when we get closer to the end of the process where the height of the product gets greater.

To see a discussion of other failure scenarios, the interested reader can consult Kharmanda (2022). When we apply the same conditions, the occurrences of failure that result happen because of uncertainty; meanwhile, when we apply different conditions, we cannot consider that the resulting occurrence of failure is due to questions of uncertainty. It can happen because of the effect of the modification of one or several starting parameters. When the starting parameters are modified, it is not necessarily the same parameters that caused the failure.

Figure 2.5. *Problem with the raft (the product can be found on 3d-printing-4u.com). For a color version of this figure, see www.iste.co.uk/elhami/uncertainty.zip*

2.6.2. *Advanced system of quality control in additive manufacturing*

Banadaki et al. (2020) proposed a quality monitoring system that can suggest corrective action by adjusting parameters in real time. The predictive quality model that they propose serves as a conceptual model for all kinds of additive manufacturing machines in order to produce reliable parts with less quality problems while limiting the waste of time and materials. Here, we propose using a monitoring system that can control quality during the manufacturing process and react by programming predictive maintenance tasks or by modifying the G-code in order to resolve the problem. For example, in Figure 2.6, we have two levels: at the first level (arrow number 1), a deviation has begun, and the extruder has modified its trajectory, while at the second level (arrow number 2), there is a scenario of total breakdown.

This type of intelligent system can be developed to indicate the performance of the process, make predictive maintenance actions possible, avoid loss of time and materials, etc. In this case, an advanced monitoring system provides a signal, and the artificial intelligence corrects this type of problem at two levels. The first level consists of executing maintenance tasks to resolve the problem and maintain a good level of quality. The remaining useful lifetime can then be evaluated. The second level consists of modifying the G-code during the process (changing certain parameters to increase the levels of adhesion and correct the trajectory). In the example presented in Figure 2.7, not only can a modification of the G-code be made, but a maintenance action (red block) can be realized, such as lubrification during the process or even the pausing (and not the stopping) of the process in order to complete another maintenance operation since the remaining useful lifetime is estimated.

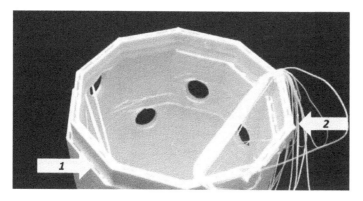

Figure 2.6. *Failure scenario at two levels (the product can be found on 3d-printing-4u.com). For a color version of this figure, see www.iste.co.uk/elhami/uncertainty.zip*

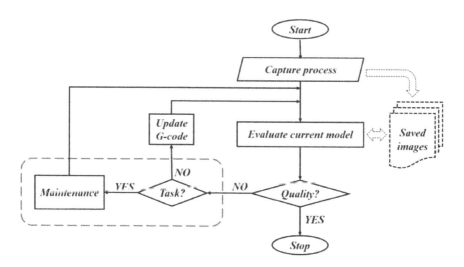

Figure 2.7. *Proposed flow chart for an advanced quality control system. For a color version of this figure, see www.iste.co.uk/elhami/uncertainty.zip*

Figure 2.8 shows a maintenance block where the process continues during the maintenance operation (e.g. lubrification). Here, it is not necessary to pause the process. However, in certain cases, we will need to interrupt the process and repair or replace certain spare parts. When the expected problem can be solved by maintenance operations, it is not necessary to modify the G-code. Several tests can be added, beyond degradation and age, as shown in Figure 2.8 (see Kharmanda et al. 2022).

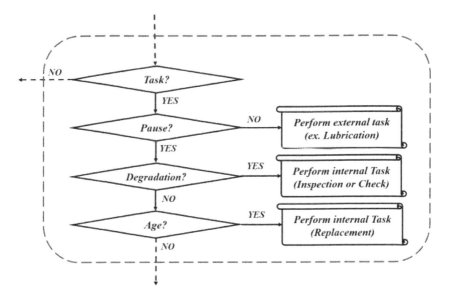

Figure 2.8. *Maintenance block. For a color version of this figure, see www.iste.co.uk/elhami/uncertainty.zip*

In the literature, other applications can be found. For example, in Ayala and López-Herrera (2021), a methodology for making multi-criteria decisions was proposed to oversee additive manufacturing equipment using preventative maintenance.

2.7. Conclusion

In additive manufacturing, several failure scenarios can be found, which require fast and judicious decision-making that accounts for the influence of uncertainty. These different failure scenarios cause additive manufacturing to be less recommendable. During certain moments of our experiences, we lost the motivation to continue using additive manufacturing at high levels. In the end, it contributed to a lower level of satisfaction, or even the full abandonment of this technology. The implementation of maintenance operations could be essential to avoid (or reduce) failure scenarios. Maintenance plays an important role in the availability of additive manufacturing machines. Thus, artificial intelligence in the case of uncertainty could be considered as an efficient tool for making decisions to reduce time and costs (reasonable time and costs). In this case, additive manufacturing could be industrialized and used to mass-produce parts.

Artificial intelligence could improve the ability to control receivers and the tasks that are executed during the additive manufacturing process, which could be very useful in reducing wasted time and materials and improving productivity levels. The use of artificial intelligence in this situation is very important from an economic point of view since it makes it possible to manage a great number of tasks (sequential, simultaneous or concurrent) and to select all compatible tasks. Thus, the effect of artificial intelligence can be considered a positive point for resolving numerous problems of uncertainty in additive manufacturing. Artificial intelligence can execute difficult and dangerous tasks for which humans require more time.

2.8. References

Alfaify, A., Saleh, M., Abdullah, F.M., Al-Ahmari, A.M. (2020). Design for additive manufacturing: A systematic review. *Sustain*, 10, 3043–3054.

Ayala, P. and López-Herrera, H.-F. (2021). Maintenance preventive analysis in additive manufacturing equipment using analytic hierarchy process. *Preprint in Research Square*. doi: 10.21203/rs.3.rs-1047953/v1.

Banadaki, Y.M., Razaviarab, N., Fekrmandi, H., Sharifi, S. (2020), Toward enabling a reliable quality monitoring system for additive manufacturing process using deep convolutional neural networks. *Material Science, ArXiv*, 6, abs/2003.08749.

Bradley, R. and Drechsler, M. (2014). Types of uncertainty. *Erkenn*, 79, 1225–1248. doi: 10.1007/s10670-013-9518-4.

Butler, K.L. (1996). An expert system-based framework for an incipient failure detection and predictive maintenance system. In *Intelligent Systems Applications to Power Systems, 1996. Proceedings, ISAP'96., International Conference on IEEE*, 321–326.

Collan, M. and Michelsen, K.-E. (2020). *Technical, Economic and Societal Effects of Manufacturing 4.0: Automation, Adaption and Manufacturing in Finland and Beyond*. Palgrave Macmillan, Cham. doi: 10.1007/978-3-030-46103-4.

Drmac, F. (2022). Reshaping organizations through artificial intelligence: Overcoming barriers of AI-implementation. Master of Sciences dissertation, Luleå University of Technology, Luleå.

Frazier, W.E. (2014). Metal additive manufacturing: A review. *Journal of Materials Engineering and Performance*, 23, 1917–1928.

Fu, Y.-F. (2020). Recent advances and future trends in exploring pareto-optimal topologies and additive manufacturing-oriented topology optimization. *Mathematical Biosciences and Engineering*, 17(5), 4631–4656. doi: 10.3934/mbe.2020255.

Gao, W., Zhang, Y., Ramanujan, D., Ramani, K., Chen, Y., Williams, C.B., Wang, C.C.L., Shin, Y.C., Zhang, S., Zavattieri, P.D. (2015). The status, challenges, and future of additive manufacturing in engineering. *Journal of Computer Aided Design*, 69, 65–89. doi: 10.1016/j.cad.2015.04.001.

Gebhardt, A. and Fateri, M. (2013). 3D printing and its applications. *RT e-Journal*, 11 [Online]. Available at: https://www.researchgate.net/publication/267433899_3D_printing_and_its_applications.

Gross, J.M. (2006). *Fundamentals of Preventive Maintenance*. AMACOM, New York.

Henry, T.C., Phillips, F.R., Cole, D.P., Garboczi, E., Haynes, R.A., Johnson, T. (2020). In situ fatigue monitoring investigation of additively manufactured maraging steel. *The International Journal of Advanced Manufacturing Technology*, 107, 3499–3510. doi: 10.1007/s00170-020-05255-4.

Janse, B. (2021). Decision making under uncertainty. *Toolshero* [Online]. Available at: https://www.toolshero.com/decision-making/decision-making-under-uncertainty/ [Accessed 2021].

Kharmanda, G. (2022). Additive manufacturing of polylactic acid (PLA) material considering preheating uncertainty effect. *Uncertainties and Reliability of Multiphysical Systems*, 22–6(1). doi: 10.21494/ISTE.OP.2022.0852.

Kharmanda, G. and Antypas, I. (2020). Reliability-based topology optimization as effective strategy for additive manufacturing: Influence study of geometry uncertainty on resulting layouts. *Journal of Physics Conference Series*, 1679. doi: 10.1088/1742-6596/1679/4/042052.

Kharmanda, G. and Mulki, H. (2022). Two decades review of reliability-based topology optimization developments. *Uncertainties and Reliability of Multiphysical Systems*, 22–6(2). doi: 10.21494/ISTE.OP.2022.0884.

Kharmanda, G., Al-Sakkaf, H., Shao, J., Bouretoua, F., Almahrji, B. (2022). NAVAIR as an effective standard of reliability centered maintenance for determining significant functional failures. *Uncertainties and Reliability of Multiphysical Systems*, 22–6(1). doi: 10.21494/ISTE.OP.2022.0849.

Li, Z. and Tsavdaridis, K.D. (2021). A review of optimised additively manufactured steel connections for modular building systems. *Industrializing Additive Manufacturing*, 1, 357–373.

Liu, J., Gaynor, A.T., Chen, S., Kang, Z., Suresh, K., Takezawa, A., Li, L., Kato, J., Tang, J., Wang, C.C.L., Cheng, X., Liang, A.C. (2018). Current and future trends in topology optimization for additive manufacturing. *Structural and Multidisciplinary Optimization*, 57, 2457–2483. doi: 10.1007/s00158-018-1994-3.

Mani, M., Lyons, K.W., Gupta, S.K. (2014). Sustainability characterization for additive manufacturing. *Journal of Research of the National Institute of Standards and Technology*, 119, 419–428. doi: 10.6028/jres.119.016.

Nakajima, S. (1988). *Introduction to TPM: Total Productive Maintenance.* Productivity Press, New York.

Poór, P., Ženíšek, D., Basl, J. (2019). Historical overview of maintenance management strategies: Development from breakdown maintenance to predictive maintenance in accordance with four industrial revolutions. In *Proceedings of the International Conference on Industrial Engineering and Operations Management*, Pilsen, July 23–26.

Pradel, P., Zhu, Z., Bibb, R., Moultrie, J. (2018). Investigation of design for additive manufacturing in professional design practice. *Journal of Engineering Design*, 29, 165–200.

Ribeiro, T.P., Bernardo, L.F.A., Andrade, J.M.A. (2021). Topology optimisation in structural steel design for additive manufacturing. *Applied Sciences*, 11, 2112. doi: 10.3390/app11052112.

Shashi, G.M., Laskar, A.R., Biswas, H., Saha, A.K.A. (2017). Brief review of additive manufacturing with applications. In *Proceedings of the 14th Global Engineering and Technology Conference*, Dhaka, 29–30 December.

Sikorska, J.Z., Hodkiewicz, M., Ma, L. (2011). Prognostic modelling options for remaining useful life estimation by industry. *Mechanical Systems and Signal Processing*, 25(5), 1803–1836.

Tebbi, O., Guérin, F., Dumon, B. (2003). Reliability testing of mechanical products – Application of statistical accelerated life testing models. In *9th International Conference on Applications of Statistics and Probability in Civil Engineering*, University of California, Berkeley.

Washington, T. (2022). Advancing AI and data science through industry/academia collaboration. *MATLAB EXPO 2022*, May 17–18.

Wiberg, A., Persson, J., Ölvander, J. (2019). Design for additive manufacturing – A review of available design methods and software. *Rapid Prototype Journal*, 25, 1080–1094.

Wong, K.V. and Hernandez, A. (2012). A review of additive manufacturing. *ISRN Mechanical Engineering*, 208760. doi: 10.5402/2012/208760.

3

Bio-Composite Structural Durability: Using Artificial Intelligence for Cluster Classification

A greater awareness of non-renewable natural resource preservation needs has led to the development of more ecological, high-performance, bio-based composite materials with new functionalities. This chapter deals with the field of structural durability in bio-based composite materials for industrial applications. The growing demand for these materials calls for an extensive investigation of their physical, chemical and mechanical behavior under different exposure conditions. The number of parameters required for characterization makes this investigation more complex. But the self-learning ability of machine learning algorithms makes this investigation more accurate and accommodates all the complex requirements.

Different damage mechanisms (transverse matrix cracking, delamination, fiber failure, interfacial debonding, debonding between the matrix and fibers) occur almost simultaneously during composite loading. This presents a scientific challenge in assigning a specific set of acoustic emission (AE) signal features to a particular damage mechanism.

It is in such a context that this chapter combines several established methods (AE measurements, feature extraction, feature selection, machine learning) and adds novelties to the feature extraction stage (deep features)

Chapter written by Abel CHEROUAT.

For a color version of all figures in this chapter, see www.iste.co.uk/elhami/uncertainty.zip.

and to the application of these methods in a unique way that provides new perspectives for the characterization of loaded composites.

3.1. Introduction

New materials have always constituted an area of research and development in civilizations and more recently in modern societies. The development of materials accelerated during the previous century with the advent of the industrial age and the exponential growth of metallurgy and steel for the manufacturing of industrial parts and massive structures. The actors in this sector have always worked to find a balance between reliability, performance and cost, making innovative materials accessible to as many people as possible. In the middle of the 20th century, synthetic materials such as plastics appeared, and they were easier to work with, lighter and less expensive. In the continued search for high-performance technical materials, the performance of low-cost metals was combined with the lightness of plastics by creating composite materials. These materials for a new age were initially thermoplastics or thermosetting polymers reinforced by fibers with far superior properties to those of the polymer on its own (Aymond and Almer 2012; Berthelot 2012).

Composites are materials with a high technical performance and advantageous physical and mechanical properties for relatively easy manufacturing and working. Over the past decade, the vision of industrial development has changed direction by integrating environmental factors into the search for innovative materials. With the objective of the technical performance having been reached, and given the question raised today by the environment and evolution of the impact of humans on their environment, a new fundamental question is raised, namely can humans continue to evolve while remaining in harmony with the environment in which they live? To respond to this issue, researchers and industry have sought to use plant resources for the manufacture of new materials. Agro-materials or bio-sourced materials now respond, in part, to this need (Wambua et al. 2003; Ashori 2008; Pickering 2008).

The context for recent research has been the mastery of raw materials and energy consumption, which remains a global objective. The lightening of mechanical structures contributes to this through the use of alternative solutions such as new, bio-sourced materials reinforced with plant fibers.

According to AGRICE (Agriculture for Chemistry and Energy), agro-materials made up of plant fibers appear to be very promising and will likely be adopted by 2030 with a provisional global growth rate of +7%. Thus, the use of plant fibers in plastics will likely amount to 320,000 tons/year, spread over the automotive, building and packaging industries.

Composite materials with natural fibers offer an interesting alternative to certain composites with usual reinforcements such as glass. Indeed, natural fibers provide advantageous mechanical properties, as well as several advantages for concerns over the environmental impact. These are renewable, biodegradable, carbon-neutral resources that require little energy to be produced. In the last few years, several studies have been conducted to have a better understanding of these fibers. The potential of natural fibers such as linen and hemp is related to the fact that their specific properties (the mechanical properties related to their volumetric mass) are comparable or even superior to those of glass fibers.

The incorporation of cellulose-based plant fibers (cotton, linen, hemp, jute, sisal, kenaf, flax, coconut, wood, etc.) in plastic, thermoplastic or thermosetting polymer materials as a replacement for glass or carbon fibers is a concept that has already been industrialized and commercialized. These composites have the potential to be the next-generation materials for structural applications in infrastructure, the automotive industry, and consumer applications of hybrid composite materials.

The reinforcement of plastic materials using plant fibers diminishes recycling problems since materials reinforced by plant fibers are easier to recycle. Polymer agro-composites and plant fibers can be ground and reinjected over several life cycles. In the future, it will be possible to eliminate agro-composites that are entirely bio-sourced as new technologies are unlocked.

Bio-based composites are an assemblage of entirely natural or partially bio-sourced materials composed of a matrix that could be a bio-polymer and plant or animal reinforcements. In order to be able to use them in various sectors such as construction, automotive, aeronautical, sports industries and others, it is imperative to know and master their physical, chemical, mechanical and environmental properties.

In order to respond to the strict demands assigned to them, these materials must notably be very resistant to the mechanical forces they are subject to (static, fatigue and shock) and under certain environmental conditions. This is particularly true with respect to car parts. When mechanical parts made up of composites with plant fibers are used in cars, these parts are subject to mechanical forces and variations in temperature and humidity, repeated throughout their use, leading to a degradation of the mechanical properties and accelerated aging of plant fibers and thus the agro-composite (Baley and Lamy 2000; Baley et al. 2004; La Mantia and Morreale 2011; Ilczyszyn et al. 2012; Antony et al. 2018, 2019).

Nevertheless, the development of composites with natural fibers in industrial applications requires more information about the degradation process in order to better predict their lifespan. Indeed, when the process of degradation begins in a composite material, a transitional wave, resulting from the liberation of stored energy, propagates from the source of the damage towards the surface of the material. AE monitoring is one of the most suitable methods to detect damage occurrence and its evolution in real time during the loading of fiber-reinforced polymers (Chen et al. 1992; Kim and Lee 1997; Kotsikos et al. 1999; Awerbuch et al. 2016).

The identification of the acoustic signature of different damage sources, such as transverse matrix cracking as the first occurring damage mechanism in composite laminates, delamination, fiber failure, interfacial debonding and debonding between the matrix and fibers, is based on an individual analysis of AE signals, which allow us to define the appearance of these critical mechanisms by identifying the scenarios where the damage happens.

Several mathematical methods exist for statistically analyzing data according to several parameters. It is indeed necessary, after collecting information, to have methods for defining the similarities and differences between the data by analyzing not only characteristic parameters, but also n parameters, also called descriptors. One of the objectives in the discriminating analysis of data is to predict whether subjects will belong to a given class based on the data by analyzing one or several variables (Berthelot and Rhazi 1990; Pappas and Kostopoulos 2001; Huguet 2002; Huguet et al. 2002; Roundi et al. 2018). Various machine learning techniques are used in combination with feature selection to extract frequency descriptors. For the automatic extraction of the intrinsic characteristics of signals, deep learning methods are becoming increasingly popular. Some commonly used deep learning approaches include auto-encoders, deep belief networks, convolutional neural networks and recurrent neural networks.

Among these techniques, we can cite analysis by principal components, the k-means or neural networks. The role of a classifier is to define the boundaries that exist between different classes. These techniques serve to identify which classes of signals, from a large overall set of signals, are coming from identical phenomena within the material, and to run tests beforehand on "textbook" samples whose behavior under stress is well known (simple matrices, unidirectional composites) (Liao 2005; Fotouhi et al. 2014; Tat et al. 2017; Zhou et al. 2018; Muir et al. 2021; Pashmforoush and Khamedi 2021).

The parameters of a classifier are estimated from a set of samples through a learning process. If the class of examples is known, the exercise then involves estimating the parameters of the discrimination function for each class. If the class of the parameter is unknown, the samples must be associated through a process of clustering or segmentation of data. The available database must be as representative as possible for it to be usable. It must be very large to find the best estimate of the parameters to the problem.

The objective of our work is to establish a practice of methods that make it possible to analyze the AE data. This is based on the use of recent classification methods (non-supervised, such as k-means, or supervised, such as neural networks). The goal of this multi-parameter statistical analysis is to elucidate the meaning of the data obtained from the AE observation of the damage caused to bio-based composite materials with polymer matrices under uniaxial traction loading (Proust et al. 2006; Farrar and Worden 2013; Hamdi et al. 2013; May et al. 2020).

3.2. The state of acoustic emission technology

Nondestructive control techniques are ways of inspecting the health or durability of materials. They make it possible to control and evaluate the health of parts after manufacture, as well as to understand their behavior when being used, and thus to predict their lifespan and plan their maintenance. X-ray radiography, ultrasonic tomography methods (acoustic emissions) and optical methods are adapted to composite materials (Touya 1979; Yuyama 1986; Godin et al. 2011; Shiino et al. 2012).

3.2.1. *Definition*

According to the AFNOR NFA 09350 norm, an AE is a phenomenon in which energy is liberated in the form of transitory elastic waves resulting from the internal, local, micro-displacements within the material under stress. This phenomenon appears within several materials when they are subject to stress of mechanical, thermal or chemical origin (Ono 1997).

The application of a charge and/or the presence of an aggressive environment produces internal modifications such as the evolution of a crack; local, plastic deformations; corrosion; and, in certain cases, phase transformations that generate acoustic emissions. One part of the energy is freed in the form of an elastic wave that spreads in all directions up to the surface of the material.

By analyzing the vibrations on the surface of the material in question, it is possible to collect, through piezo-electrical means, information about the event that caused these vibrations and created the AE signal. On each receiver, the activity (the number), intensity, power and duration are analyzed in real time. The detection, acquisition and analysis chain include (Figure 3.1):

– a piezoelectrical receiver for which resonance frequencies are captured between 100 kHz and 1 MHz;

– a preamplifier of the signals captured by the receiver whose amplification varies between 20 and 80 dB to eliminate the mechanical noises of parasitic environments and noise.

We can then distinguish two types of emissions:

– Discreet acoustic emissions: the AE signal or the burst appears to be a *damped sine wave*. In discrete AEs, transitory signals, or bursts, are measured. If the bursts become too frequent with respect to their duration, they overlap and then become impossible to separate. This is principally observed in **composite materials**.

– Continuous acoustic emissions: the AE signal appears as an *augmentation in background noise*. This is principally observed during plastic deformation in **metallic materials**.

With respect to acoustic and vibratory methods, the AE technique provides two advantages:

– better sensitivity due to the fact that environmental noise is greatly attenuated by ultrasonic frequencies;

– better temporal resolution for discerning brief impulsions (one impulsion = one implosion).

3.2.2. *Applications of acoustic emissions*

Measuring acoustic emissions is a non-intrusive and nondestructive method for identifying and monitoring the evolution of numerous systems, whether they undergo mechanical stress or not. AE thus provides numerous advantages with respect to the other nondestructive control techniques (NDC). Thanks to AE, it is possible to detect the presence of an error as it occurs, localize it and determine its nature and gravity. However, this technique is incapable of detecting errors if they do not evolve, nor can it measure them. AE has a high sensitivity to detect phenomena that other NDC techniques cannot detect. The target applications in industry are:

– monitoring technical construction in progress;

– detecting leaks in industrial structures;

– monitoring manufacturing quality;

– identifying the durability of materials;

– preventive conditional maintenance;

– measuring auditive health;

– monitoring structural durability.

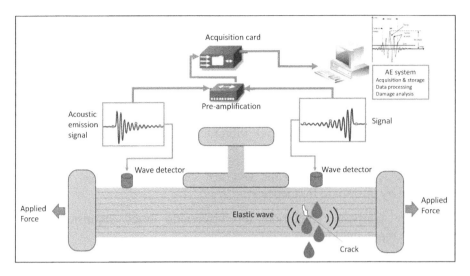

Figure 3.1. *Principle of acoustic emission*

3.2.3. *Mechanisms responsible for AE in materials*

The mechanisms that are responsible for AE are of a different nature in different materials. When a material undergoes stress, the mechanisms of the damage caused by this stress generate AE.

– The mechanisms responsible for AE in metals:

One of the potentially emissive mechanisms in terms of acoustic emissions in metals is the movement of dislocations. The recorded signals can provide an image of the implicit relaxation of energy during:

1) dislocation movements;

2) creation and spread of cracks;

3) rupture or decohesion of inclusions;

4) phase transformations in martensite.

– The mechanisms responsible for AE in concrete:

The advantage of the AE technique when applied to concrete structures is the localization of internal types of damage:

1) creation of micro-fissures in granules (gravel);

2) creation of micro-fissures in the matrix (cement);

3) creation of micro-fissures in the granule/matrix interface;

4) growth of already existing fissures.

– The mechanisms responsible for AE in composite materials:

The application of a mechanical load onto a composite material leads to the initiation of discontinuities on the surface (fissures) or in the volume (micro-cavities) and to the creation of damage that will develop primarily in certain directions. The initiation and propagation of these local discontinuities lead to the relaxation of the energy stored within during the application of the stress, creating elastic waves that will spread through the material. The different mechanisms causing damage within these materials are principally (Figure 3.2):

1) transverse matrix cracking delamination;

2) fiber failure;

3) interfacial debonding;

4) debonding between the matrix and fibers;

5) fiber pulling.

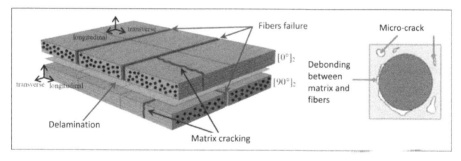

Figure 3.2. *Mechanisms responsible for damage in composites*

3.2.4. *Propagation of waves through materials*

A wave corresponds to the propagation (in space and time) of a perturbance, producing with its passage a reversible variation of local, physical properties without the displacement of the material. Ultrasonic waves are waves of vibration that spread within elastic materials at speeds that will vary according to the environment and the type of wave. If the speed of propagation of the perturbance is perpendicular to the vibratory speed, the wave is transversal (vibration in a cord) (Dieulesaint and Royer 1974; Scruby 1987; Ohtsu and Ono 1988; Benzeggagh et al. 1992; Ceysson et al. 1996; Scida et al. 2002; Khamedi et al. 2020; Saeedifar and Zarouchas 2020).

Elastic waves break down into body waves and surface waves. The two types of body waves used in nondestructive control are longitudinal waves and transversal waves. Longitudinal waves, called compression or dilation waves, are faster, as they are the first to be detected by receivers (Gorman and Prosser 1991; Keprt and Benes 2008).

If a small strain, denoted as ε, is applied to an elastic solid, the stress produced is $\sigma = E. \varepsilon$ (E is the elastic modulus) relaxes and releases transitory elastic waves. The fundamental principle of dynamics applied to a voluminous element (in the one-dimensional case) can be written according to the following Alembert equation as:

$$\begin{cases} \text{div}(\sigma) = \dfrac{\partial\sigma}{\partial x} = E\dfrac{\partial\varepsilon}{\partial x} = E\dfrac{\partial}{\partial x}\left(\dfrac{\partial u}{\partial x}\right) = E\dfrac{\partial^2 u}{\partial x^2} = \rho\dfrac{\partial^2 u}{\partial t^2} \\ \dfrac{\partial^2 u(x,t)}{\partial x^2} - \dfrac{1}{v_L^2}\dfrac{\partial^2 u(x,t)}{\partial t^2} = 0 \end{cases}$$

[3.1]

where ρ is the volumetric mass, u(x,t) is the displacement of the material within the solid and $s_L = \sqrt{\dfrac{E}{\rho}}$ is the longitudinal speed of propagation. The transversal speed of propagation, as a function of the Poisson coefficient of the material v, is equivalent to: $s_T = \sqrt{\dfrac{E}{2\rho(1+v)}}$.

We should note that the speed of propagation becomes larger as the environment becomes more rigid (large Young's modulus) and less inert (low volumetric mass). The decrease in the speed of the wave depends on the decrease in the elastic modulus of the material (the damage caused). With a modal approach, the localization of the sources of the damage can be found in a rather precise way as a function of the degradation of the rigidity modulus.

For isotropic materials (the 3D case), the excitation of the environment by a source brings about the propagation of two types of transitory mechanical waves that propagate in concentric spheres centered on the source. First, a longitudinal wave moves at the following speed:

$$S_L = \sqrt{\frac{E(1-v)}{\rho(1+v)(1-2v)}}$$

[3.2]

The solution to the Alembert differential equation is:

$$\frac{\partial^2 u(x,t)}{\partial x^2} = \frac{1}{c^2}\frac{\partial^2 u(x,t)}{\partial t^2}$$

[3.3]

In order to solve the wave equation, we continue with the following change in variables:

$$\begin{cases} \alpha = t - x/c \\ \beta = t + x/c \end{cases} \Rightarrow \begin{cases} \dfrac{\partial \alpha}{\partial t} = 1 \text{ and } \dfrac{\partial \alpha}{\partial x} = -\dfrac{1}{c} \\ \dfrac{\partial \beta}{\partial t} = 1 \text{ and } \dfrac{\partial \beta}{\partial x} = \dfrac{1}{c} \end{cases} \Rightarrow \begin{cases} t = \dfrac{(\alpha+\beta)}{2} \\ x = c\dfrac{(\beta-\alpha)}{2} \end{cases}$$

[3.4]

We can then express the function u in terms of the variables α and β and the partial derivatives of x and t:

$$\begin{cases} \dfrac{\partial u}{\partial t} = \dfrac{\partial u}{\partial \alpha}\dfrac{\partial \alpha}{\partial t} + \dfrac{\partial u}{\partial \beta}\dfrac{\partial \beta}{\partial t} = \dfrac{\partial u}{\partial \alpha} + \dfrac{\partial u}{\partial \beta} \\ \dfrac{\partial u}{\partial x} = \dfrac{\partial u}{\partial \alpha}\dfrac{\partial \alpha}{\partial x} + \dfrac{\partial u}{\partial \beta}\dfrac{\partial \beta}{\partial x} = \dfrac{1}{c}\left(\dfrac{\partial u}{\partial \beta} - \dfrac{\partial u}{\partial \alpha}\right) \end{cases} \Rightarrow \begin{cases} \dfrac{\partial}{\partial t} = \dfrac{\partial}{\partial \alpha} + \dfrac{\partial}{\partial \beta} \\ \dfrac{\partial u}{\partial x} = \dfrac{1}{c}\left(\dfrac{\partial}{\partial \beta} - \dfrac{\partial}{\partial \alpha}\right) \end{cases}$$

[3.5]

Applying the derived operators a second time:

$$\begin{cases} \dfrac{\partial^2 u}{\partial x^2} = \dfrac{1}{c^2}\left(\dfrac{\partial^2 u}{\partial \alpha^2} - 2\dfrac{\partial^2 u}{\partial \alpha \partial \beta} + \dfrac{\partial^2 u}{\partial \beta^2}\right) \\ \dfrac{\partial^2 u}{\partial t^2} = \dfrac{\partial^2 u}{\partial \alpha^2} + 2\dfrac{\partial^2 u}{\partial \alpha \partial \beta} + \dfrac{\partial^2 u}{\partial \beta^2} \end{cases}$$

[3.6]

and inserting these partial derivatives in the wave equation [3.3], we obtain:

$$\frac{1}{c^2}\left(\frac{\partial^2 u}{\partial \alpha^2} - 2\frac{\partial^2 u}{\partial \alpha \partial \beta} + \frac{\partial^2 u}{\partial \beta^2}\right) - \frac{1}{c^2}\left(\frac{\partial^2 u}{\partial \alpha^2} + 2\frac{\partial^2 u}{\partial \alpha \partial \beta} + \frac{\partial^2 u}{\partial \beta^2}\right) = \frac{\partial u}{\partial \alpha}\left(\frac{\partial u}{\partial \beta}\right) = 0 \qquad [3.7]$$

This means that for the function u(x,t) to be a solution, the function $\partial u/\partial \beta$ must not depend on the variable α, which gives:

$$\frac{\partial u}{\partial \beta} = \gamma(\beta) \Rightarrow \frac{\partial u}{\partial \alpha}\left(\frac{\partial u}{\partial \beta}\right) = u(\alpha,\beta) = f(\alpha) + g(\beta) \qquad g(\beta) = \int \gamma(\beta) \ [3.8]$$

It is possible to integrate $\partial u/\partial \beta$ with respect to the variable β to find the expression of the function u(x,t). The variables α and β are now separated. The general solution to the Alembert equation is the sum of two progressive waves advancing in inverse directions (incident waves called dilation or compression waves) with a speed of:

$$s_L = \sqrt{\frac{E}{2\rho}} \qquad \text{and a reflected or rotating wave with a speed of}$$

$$s_T = \sqrt{\frac{E}{2\rho(1+v)}} \ ).$$

The general solution can be written as:

$$u(x,t) = \underbrace{f\left(t - \frac{x}{c}\right)}_{\text{progressive wave}} + \underbrace{g\left(t + \frac{x}{c}\right)}_{\text{progressive wave}} \qquad [3.9]$$

where *f* and *g* are propagation functions of a wave progressing in the + and – directions, respectively.

For a harmonic or sinusoidal progressive wave: this is a superposition of a signal propagating towards the right at a speed of +c and another signal propagating towards the left at a speed of –c. If the frequency is the product of its wave number and the speed, the solution is:

$$u(x,t) = A\cos(kx - \omega t + \varphi) \qquad [3.10]$$

where *A* is the amplitude, *k* is the wave number, *ω* is the angular natural frequency and *φ* is the frequency phase; the period $T = 2\pi/\omega$, the length of the wave (spatial period) $\lambda = 2\pi k$ and the natural frequency $f = 1/T$.

For a stationary wave: this is a superposition of two progressive waves with the same amplitude *A* and the same wave number *k*, $\omega = k.c$ propagating at the respective speeds of +*c* and −*c*:

$$u(x,t) = A\cos(kx - \omega t) + A\cos(kx + \omega t) = 2A\cos(kx + \varphi).\cos(\omega t + \phi) \quad [3.11]$$

where the displacement $u(x,t)$ is characterized by the product of two terms that depend on the spatial variable for one and the temporal variable for the other.

This wave is stationary because it does not propagate: the spatial dependency of $2A\cos(kx + \varphi)$ is modulated in its amplitude by a function that depends on time, $\cos(\omega t + \phi)$. The transfer of energy in the direction of the progressive wave is compensated for by the transfer in the direction of the retrograde wave: overall, therefore, there is no energy transfer.

There are places whose vibrational state is null at any given time (destructive interference): these are the *nodes*, which are at a distance of λ\2. The *troughs* correspond to the places where the deformation is maximal (constructive interference), but depend on time in a harmonic way (Figure 3.3).

Detecting and interpreting AE signals remains a very delicate task when dealing with complex structures or very large ones. We know that AE translates into the propagation of mechanical waves that are most often surface waves if the structure is thick enough. These waves radiate from the source and their propagation speed is constant, whatever the vibration frequency may be, and its value depends on the nature of the control material. The localization of an AE source from a surface requires the use of receivers. Understanding AE mechanisms requires the statistical analysis of these impulses. Therefore, it is possible to detect the appearance of a particular source during use or to eliminate undesirable or parasitic signals.

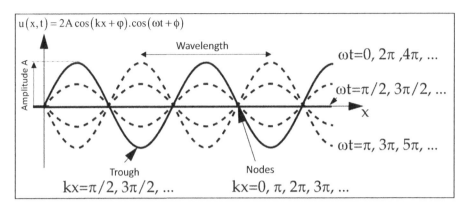

Figure 3.3. *Propagation of a stationary mechanical wave through materials*

3.2.5. *Phenomenological approach to acoustic emissions*

Analyzing AE signals makes it possible to associate signals with the source of a mechanism. In phenomenological AE, each recorded signal has properties related to the characteristics of the source. The two parameters that influence the signal are:

– the acquisition threshold, which represents the minimal amplitude for which a signal will be recorded;

– the acquisition speed, since an event or a burst is recorded only when its amplitude surpasses the value of the acquisition threshold.

Descriptors are calculated over the temporal form and the frequential spectrum (calculated with a Fournier transformation) of measured signals. The parameters (Figure 3.4) that make it possible to characterize a type of signal in order to identify the different mechanisms at play are presented in Table 3.1.

Parameter	Unit	Formula	Definition
Amplitude peak	Decibel, dB	$A = 20\log_{10}\left(\dfrac{V_{max}}{1\mu V}\right)$	Maximal amplitude of the signal throughout the burst duration
Duration	Microsecond	T	The time separating the first and last time the threshold is passed
Number of beats		NC	Number of times the threshold is passed by the signal over its entire duration
Time to peak	Microsecond	TM	The time separating the first passing of the threshold and the peak amplitude of the signal
Average frequency	kHz	$FM = \dfrac{NC}{T}$	Number of beats in a burst divided by its duration
Signal energy	Attojoules	$EA = \int A^2 dt$	Integral of the square of the burst's amplitude over the entire duration of the signal

Table 3.1. *Parameters defining an AE burst*

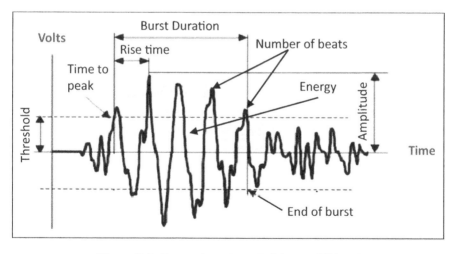

Figure 3.4. *Parameters characterizing an AE burst*

3.3. The concept behind classification

3.3.1. *Introduction*

A classification is a set of algorithms used to discriminate between the events associated with the various AE sources they originate from using classification rules created from the study of such source events that have been identified and associated with mechanisms causing micro-fissures.

Several classification algorithms are available in the literature and accessible via different statistical software based on linear or quadratic discriminating functions or the evaluation of the K-nearest neighbor. The identification of the mechanisms causing damage based on their acoustic signature is done according to the following steps:

1) collection of data through experimentation using mechanical tests on composites;

2) extraction of vectors of significant characteristics that allow us to separate AE signals;

3) selection of AE descriptors or parameters to be used;

4) regrouping collected signals by class via a non-supervised classification method based on identified descriptors;

5) choice of classifiers and optimization;

6) labeling of classes through the correlation between identified classes and corresponding damage mechanisms;

7) construction of the learning database.

3.3.2. *Statistical analysis in the identification of different source mechanisms*

– *Choice of descriptors*:

In AE, before applying classification rules, the information contained in these must first be reduced into a simpler form. Each AE signal can be described according to several parameters (amplitude, energy, duration, time to peak, average frequency, etc.). It is therefore necessary to:

- judiciously choose these descriptors;

- optimize the number so as to not lose information about the signals, without having so many as to incumber the calculations or skew the relative importance of each parameter.

– Exploration of data:

The exploration of data in general terms means to extract or round out the data, which appears in different forms, in order to obtain models and gain knowledge about this model. In the data exploration process, large sets of data are first sorted, then models are identified and relationships are established so that data analysis can be performed and problems can be resolved.

– Processing AE signals:

After the extraction of parameters for each of the AE signal classes, the classes for which the parameters are known must be defined. For our AE analysis system, these classes correspond to the classes of AE signals that have been associated with a process of fissure creation (matrix, fiber, fiber–matrix interface, peeling between folds).

– Classification methods:

These require discriminating descriptive analysis to make a decision about the classification of a signal belonging to a group of unclassified signals with respect to a group of classified signals in previously defined classes. The result is given in the form of the percentage of classified signals in another class from the one these belong to. This value, which we call the classification error rate, is the principal criterion for comparison of the discriminating power of a parameter set for which a given discrimination function has been applied.

– Learning classifiers (machine learning):

The algorithms for machine learning are numerous. A good machine learning model is a model that generalizes. Generalization is the capacity of a model to make prediction not only regarding the data that has been used to build it, but also and especially about new data: this is the learning. The most commonly used classification algorithms include:

1) probabilistic methods: Bayes classifier;

2) methods based on examples: K-nearest neighbor;

3) methods based on attributes: flowcharts, neural networks, linear regression, logistical regression, genetic algorithms and the Markov decision process (Oja 1989; Schalkoff 1997; Adams and Hand 1999; Dubuisson 2001; Zhang and Friedrich 2003; Honeine and Richard 2007; Marec et al. 2008; Alpaydin 2009; Hand 2009; Harizi 2012; Bajaj and Bilas 2013; Fukunaga 2013; Chaki et al. 2016).

3.3.3. *Classification method*

Training data:

$X = \{(x_i, y_i)/i = 1, ..., N\}$ represented by a vector of dimension m.

1) Each data point x_i is characterized by m attributes and its class $y_i \in Y$:

– Y is discrete;

– $|Y| = 1$: mono-class classification;

– $|Y| = 2$: binary classification;

– $|Y| > 2$: multi-class classification.

2) The data must be chosen carefully:

– sufficient in number;

– covering different categories;

– not distorted by noise.

A classifier is constructed by analyzing a training database with its respective classes. There are two distinct methods of classification: supervised and unsupervised.

3.3.3.1. *Supervised classification*

If we have a library of already identified and classified signals, **supervised** classification of new signals is done by comparison with the signals in the library: in this case, the classes are known a priori. We have different learning strategies.

Discriminating approaches are based on a probability model:

– logistical regression;

– flow charts;

– neural networks.

Generative approaches are based on a statistical model:

– linear analysis;

– K-nearest neighbors.

To do this, a classifier must be created. This is a sort of black box where information will be submitted about the object and the corresponding class of the object will be provided. For this, the classifier must be a function or a mathematical model that will produce the closest result possible to the one desired when entries are submitted. But it is impossible to construct a generalist function that would work for all entries and all results. A function must thus be constructed using the entry and result data from examples or samples that are known for the problem. This procedure unfolds as follows:

– **Learning phase**: the examples are submitted one by one in a random order to the classifier. If the classifier finds the correct response, we move to the next one, and if the classifier does not find the response, in this case, the training algorithm repairs the classifier so that it provides the correct response.

– **Test phase**: after this so-called training phase of the classifier, there is a test phase. During this phase, examples that have never been seen by the classifier during its training are submitted to it. If the responses are correct, the classifier has been appropriately trained, and on the contrary, if they are incorrect, the classifier has been badly trained and it is then important to find the cause. Several scenarios are possible:

1) lack of data: the training data was not sufficiently descriptive, with the consequence that it is impossible to find the corresponding class;

2) over-learning: the classifier is so well trained on the learning data that it consequently makes errors on new data;

3) divergence: if the training database is not representative of reality, then the learning database diverges too much from the test database;

4) under-training: the training was stopped in the middle of iterations, so it did not have enough time to adapt to the problem (the best solution is to add iterations during the training of the classifier).

K-nearest neighbors (k-NN)

The "k-nearest neighbors" (k-NN) method is one of the simplest supervised learning methods. Starting with a cluster of points for which the classes are known, how do we classify a new point for which the class is unknown? The principle of the method is to assign new, unclassified objects to the class to which the majority of its k-nearest neighbors belong by calculating the distances:

$$d(x,y) = \sqrt{\sum_{i=1}^{n}(x_i - y_2)^2} \qquad [3.12]$$

An example of the *k*-NN classification (Figure 3.5):

– if point number $n = 3$ (circle made of a continuous line), it is assigned to class 2 (two triangles and only one square) in the circle in question;

– if $n = 5$ (circle made of a dotted line), it is assigned to class 1 (three squares in front of two triangles in the external circle);

– if $n = 10$ (large circle), it is assigned to class 1 (five squares in front of four triangles in the external circle).

How do we choose the appropriate K (the number of neighbors to take into account)?

– if K is too small, the classification becomes too sensitive to noise;

– inversely, if K is too large, the classification can be wrong due to an overly large decision zone.

The choice of the K class is therefore dependent on the input data:

– if there is a *large dispersion* (large gap-type) → K "large";

– if the data points are *overlapping* and the point being classified is *in the cluster* → K "mean";

– if the point being classified is *outside of the cluster,* we can choose → K "small".

Should we use a number of high, representative points n for the learning set?

– If we use a very large number of representative points:

- the learning is more precise and minimizes classification errors;

- however, it is also much slower → therefore, no.

– If we use too low of a number of representative points:

- the learning contains errors and heuristic additions are necessary;

- however, the learning is very fast.

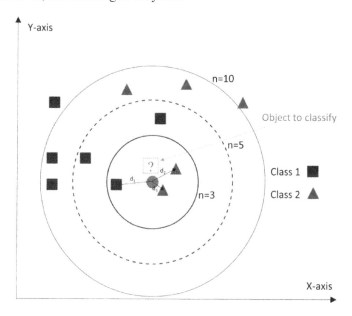

Figure 3.5. *Diagram of the principle of the k-NN method with a classification problem in two classes*

3.3.3.2. *Unsupervised classification*

If we do not have access to already identified or classified signals, classes or groups of individuals with common characteristics must be identified: in this case, the number and definition of the classes are not a given a priori. The objective is to group the data according to resemblance.

The task of the user is to identify the obtained subsets, that is, the identification of the physical phenomenon that each subset characterizes. There are **two principal methods** of unsupervised learning:

– partitioning methods such as with the algorithms for k-means;

– hierarchical grouping methods.

Partitioning methods: algorithms for K-means

The k-nearest neighbor method consists of seeking the example that is closest to the one that is being analyzed within a database and then providing the user with the solution to this example as a response. The quantity of K classes we wish to find must first be determined by making sure to **minimize intra-class inertia** (Figure 3.6) according to the following procedure:

1) The K class centroids are placed randomly.

2) Each point in the cluster is taken and **assigned to the class of the centroid** that it is closest to. We thus obtain a cluster of n points (x_1, x_2, ..., x_n) for which each point corresponds to one of the K classes:

$$\arg\min \sum_{i=1}^{k} \sum_{x_j \in S_i} \left\| x_j - C_i \right\|^2 \qquad [3.13]$$

3) We **calculate the center of gravity** for each class:

$$C_k = \frac{1}{m_k} \sum_{x \in C_k} x_i \qquad [3.14]$$

4) We **move the centroid** onto the newly calculated center of gravity.

5) We **recalculate**, for each of the points in the cluster, which is the closest centroid, in order to assign it to the correct class.

6) We recalculate the new center of gravity, and so on and so forth until we reach the **convergence of the algorithm** (when **nothing moves** from one iteration to the next, i.e. when the centroid remains immobile, even after recalculating the class for each point).

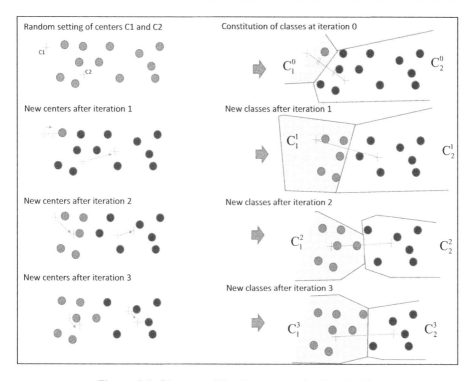

Figure 3.6. *Diagram of the K-mean method's principle
for a two-class classification problem*

– Disadvantages of the k-means algorithm: the shortcomings of the classification method by the k-means algorithm are:

1) K must be fixed;

2) the result depends greatly on the selection of initial centers. It does not necessarily provide the optimum result; it provides a local optimum that depends on the initial centers.

Solution: to select the optimal number of classes, the K-means is often started several times, with different values of K. For each one, we make note of the **intra-class inertia** that is obtained. The more classes there are, the more the number of individuals per class goes down, and the more the classes become tight.

To determine the number of classes to study, we study the graph that traces intra-class inertia as a function of the number of classes. We are especially seeking a "break" in the curve. This "break" indicates the point at which the number of classes has grown too big. We see that inertia diminishes very quickly between two and three clusters, and more slowly between eight and nine clusters. Thus, we can conclude that the optimal number of classes is 3 (see Figure 3.7).

– Other versions of the k-means algorithm exist:

1) global k-means;

2) detecting misclassification errors;

3) incremental approach (or modified fast global K-means).

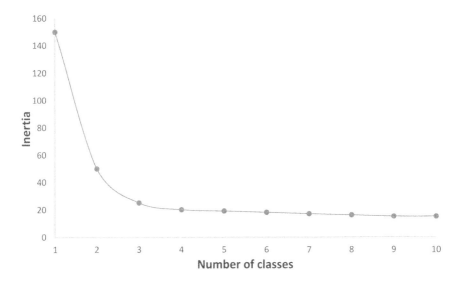

Figure 3.7. *Choosing the optimal number of classes*

3.4. Application to the classification of damage mechanisms in agro-composites

An experimental study of the different damage mechanisms of eco-composite materials made of hemp fibers and polypropylene (PP) resin was done. The acoustic signals were recorded during the traction trials on test tubes made by heated compression following different fiber orientations in order to privilege certain modes of damage. Acoustic signatures coming from four modes of damage were identified: matrix fissure formation, fiber–matrix decohesion, peeling and fiber breakage. Multi-variable (k-means) statistical analysis techniques were used to distinguish the corresponding signals from the damage modes through an AE data classification approach (amplitude, frequency, burst number) (Tabrizi et al. 2019; Šofer et al. 2021).

3.4.1. *Fabrication of agro-composite test tubes*

In this study, we made unidirectional test tubes from a PP/hemp composite in different orientations: 0°, 45°, 67.5° and 90°, as well as those that were stratified according to the following protocol (Figure 3.8):

1) The reinforcements made of dried hemp fabric were cut to the required dimensions and dried in the oven for 30 minutes at a temperature of 160°C in order to avoid humidity, which can provoke air bubbles in composites.

2) The fabric pieces were placed in a pile between two sheets of PP in the desired orientation.

3) A closed mold was placed in the oven for 2 hours and 30 minutes at a temperature of 190°C, which allows PP composites to melt (stages 1 and 2) and ensures uniform distribution of the matrix.

4) The mechanical press was preheated for 1 hour and 40 minutes at a temperature of 190°C (stage 3). The composites were heat-pressed at 6 bars for 10 minutes.

5) The mold was allowed to cool (0.95°C/min); the composite sheets were removed from the mold (stage 4).

6) The final composite sheets with a reinforced PP matrix were cut to a width of 200 mm and a length of 200 mm.

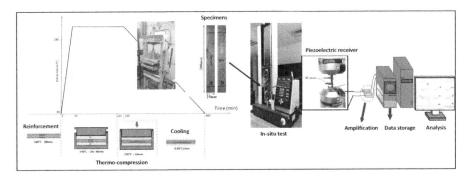

Figure 3.8. *Forming procedure of bio-based composite sheets reinforced by hemp fibers*

3.4.2. *Characterization of tensile tests*

Different modes of the classification of AE signals have been created using a set of signals from identified and non-identified sources. Each of these classes is associated with a rupture phenomenon in the different materials constituting a stratified composite. Four classes of test tubes were prepared in this study and each one of them was put under stress in order to produce either peeling ruptures in the stratified test tubes or traction ruptures in the test tubes (see Figure 3.9):

1) A traction test on a composite test tube with its orientation at [0°] with respect to the loading direction gives the class associated with **fiber ruptures**.

2) A traction test at [+/-45°] is used to identify the class associated with a **matrix crack**.

3) A traction test at [+/-67.5°] is used to identify the class associated with **fiber–matrix debonding**.

4) For a test at [90°], most of the expected damage is in the class of **peeling at the folds**.

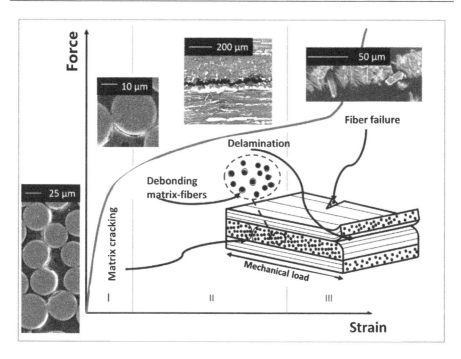

Figure 3.9. *Tensile tests: stress–strain evolution and damage mechanisms in agro-composites*

Traction tests on the test tubes were done at room temperature, or 25°C, with a traction machine made by MTS. The speed of displacement of the specimen was fixed at 0.1 mm/min. The standardized traction test tubes (with the following dimensions: length = 100 mm, width 10 = mm and thickness = 3 mm) were performed in three directions with respect to the axis of the fibers for the unidirectional composite: 90°, 67.5° and 45°.

For the creation of the learning database, a test at [0°] is used to identify part of the class associated with the rupture of fibers, a test on the matrix or the fibers at [+/-45°] is used to create the part of the class associated with a rupture in the matrix, and finally, a test at [+/-67.5°] is used for fiber–matrix debonding.

3.4.2.1. *Test on a polymer matrix*

Traction tests at room temperature were performed on a resin plate to determine the acoustic signature of the source mechanism corresponding to the formation and growth of damage within this matrix. The constraint/deformation curve following the acoustic activity for a traction test on only resin is presented in Figure 3.10. The evolution of the bursts as a function of the amplitude of the signal is presented in Figure 3.11. The number of accumulated beats represents acoustic activity:

– Phase 1: strain from 0 to 2.5% → negligible damage and no signal detected.

– Phase 2: strain from 2.5% to 7.5% → initiation of deformation: acoustic activity characterized by signals whose amplitudes are between 35 and 50 dB. The signals with these characteristics are referred to as type A.

– Phase 3: strain > 7.5% → rupture: the signals have a weak amplitude (between 40 and 65 dB) and very weak energy, and cannot be used since they are associated with the rupture of the PP matrix and not its progressive damage.

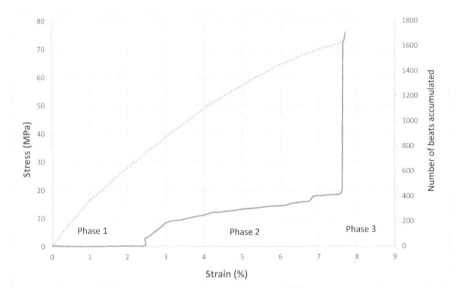

Figure 3.10. *Stress–strain curve following acoustic activity in a polypropylene matrix*

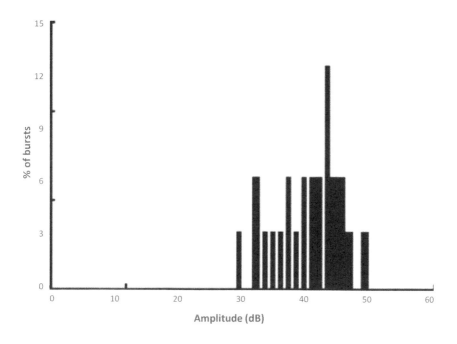

Figure 3.11. *Amplitude distribution of the polypropylene matrix under tensile stress test*

3.4.2.2. *Tensile test on a composite [0°]*

Damage initiation mechanisms, when static, are principally associated with the matrix. Indeed, the spread of the amplitude as a function of time demonstrates the presence of three signal classes: the first corresponds to the appearance of fissures in the matrix (*type A*), followed by fiber–matrix decohesion (*type B*), then intra-lamination decohesion or peeling (*type D*), and finally, the rupture of fibers (*type C*).

Figure 3.12 provides the constraint/deformation curves derived from traction tests. The "macroscopic" mechanical behavior of agro-composites is practically linear until the test tubes, which are fragile, rupture. At 0°, a composite has a "fragile, elastic" behavior whose rigidity and resistance are proportional to the number of folds at 0°. The mechanical properties are presented in Table 3.2.

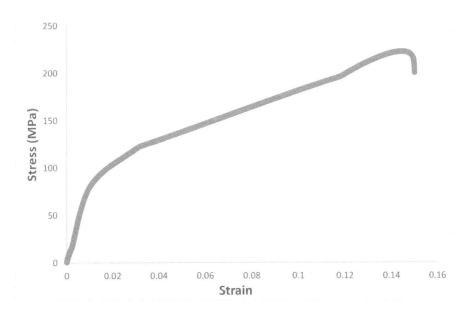

Figure 3.12. *Stress–strain of tensile tests on unidirectional fiber at 0°*

We considered that the average absolute energy characterizes the phases of damage and rupture for different types of composite materials. The energy was calculated using each energy characteristic for wave forms. Then, we combined the energy samples to find the nonlinear cumulative energy curve from the lowest to the highest in chronological order. The significant changes in the gradient of this cumulative energy curve are considered the moment that damage occurs in the region of the defect.

The samples of AE data were considered for the observation of agro-composite behavior as a function of the signal's energy. Figure 3.13 shows the correlation of data between the AE data and the mechanical data for an orientation of 0°. It presents the initiation of different phases in the gradients of cumulative energy curves with the increment of the mechanical load. From this, we can affirm that the AE data has a strong correlation with the damage caused to composites.

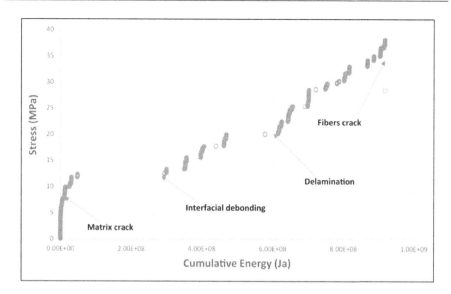

Figure 3.13. *Identification of damage as a function of cumulative energy*

3.4.2.3. *Tensile test on composites [+/-45°]*

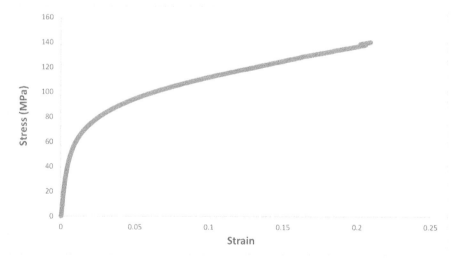

Figure 3.14. *Stress–strain curve for tensile tests at [+/-45°]*

Tensile tests were performed in the same conditions as before on a unidirectional composite oriented at 45° with respect to the axis of the fibers. Figure 3.14 presents the constraint/deformation curve following acoustic activity (cumulative number of beats) obtained for one of these tests. As Figure 3.15 indicates, acoustic activity becomes significant after a deformation of 5%. It then increases regularly until the sample reaches the rupture point.

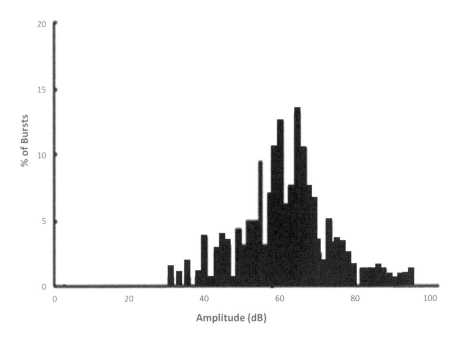

Figure 3.15. *Amplitude distribution of characteristic signals for acoustic emissions collected from unidirectional composites with perpendicular stress to fiber orientation*

3.4.2.4. *Tensile test on composites [+/-67.5°]*

Tensile tests were performed in the same conditions as before on the unidirectional composite in a direction oriented at 67.5° with respect to the axis of the fibers, and once again, 80% of the signals were assigned to the principal damage expected: decohesion. The samples presented a fragility characterized by a relatively straight constraint/deformation curve and a weak rupture deformation (0.07). The rupture constraint was 40 MPa. Figure 3.16 presents the constraint/deformation curve following the acoustic activity (cumulative number of beats) obtained for one of these tests. As the figure indicates, acoustic activity becomes significant after a deformation of 5%. It then rises regularly to the sample's point of rupture.

In the study, the average absolute energy was used to characterize the phases of damage and rupture of the different types of composite materials. The significant changes in the gradient of this cumulative energy curve are considered to be the time of damage to the region. Figure 3.17 shows the correlation of data between AE data and mechanical data for orientations of [+/-67.5°]. It presents the initiation of different phases in gradients of the cumulative energy curves with the increment for mechanical load.

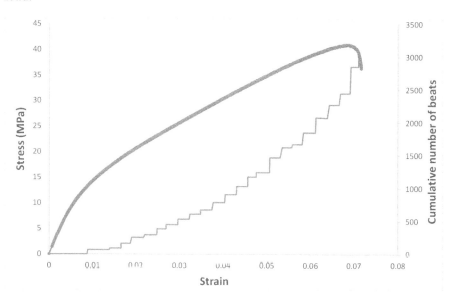

Figure 3.16. *Constraint/deformation curve following acoustic activity of composites at [+/-67.5°]*

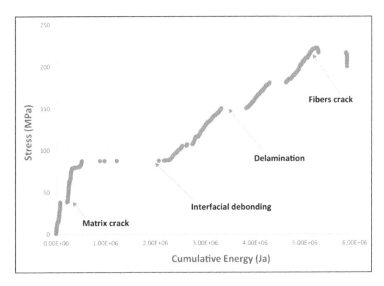

Figure 3.17. *Identification of damage to composites [+/-67.5°] as a function of the signal's cumulative energy*

3.4.2.5. *Tensile test on composites at [90°]*

The samples display a fragility characterized by a relatively straight stress/strain curve and a strain with a weak rupture point (0.1). The rupture constraint is 35 MPa. Tensile tests were performed in the same conditions as before on the unidirectional composite in a direction oriented at 90° with respect to the axis of the fibers. The amplitude distributions of bursts received during traction tests at 90° with respect to the axis of the fibers are presented in Figure 3.18. For these tests, most of the acoustic emissions received are in the same range of amplitudes as the results obtained from resin alone. Numerous signals with greater amplitude were also detected (70–90 dB). In Figure 3.19, the chronology of the appearance of these signals as a function of their amplitude shows that those with large amplitudes only appear at the end of the test. They are therefore associated with a consecutive source phenomenon of peeling between layers.

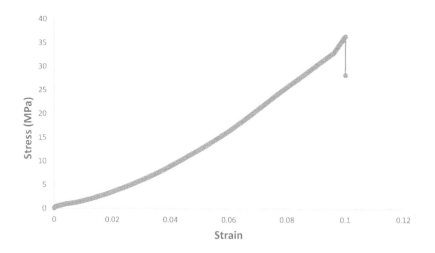

Figure 3.18. *Stress–strain curve for unidirectional composites tested perpendicularly to fibers*

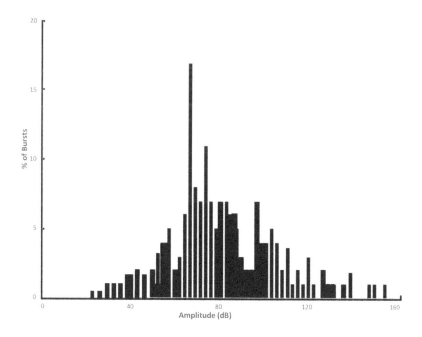

Figure 3.19. *Amplitude distribution of characteristic acoustic emission signals for unidirectional composites tested perpendicularly to fibers*

	[0°]	[+/-45°]	[+/-67.5°]	[90°]
Maximal stress	250 MPa	140 MPa	40 MPa	35 MPa
Strain at rupture	15%	22%	7%	10%

Table 3.2. *Maximal stress and strain at rupture of agro-composites*

3.4.3. *Classification of damage mechanisms*

– The damage caused to the test tubes in traction was monitored using an acquisition system (EPA) containing two tracks with one sampling a frequency of 5 MHz, one LDLC acquisition card connected to a microcomputer and two Micro-80 piezoelectric receivers with a bandwidth of 100 kHz–1 MHz. The signals were then amplified by a preamplifier with an amplification of approximately 40 dB.

– AE bursts were acquired using a MISTRAS defect identification system.

– Noesis software, specialized in analysis and classification of AE data in post-processing, as well as in real time, was used following the selection of five of the most relevant parameters: amplitude, energy, time to peak, duration and number of beats.

– The classification allowed us to identify the damage modes presented in Table 3.3.

Type of damage	Amplitude A (dB)	Frequency (kHz)	Time to peak TP (µs)	Duration T (µs)
Matrix crack (A)	35–65 Average	90–180	> 0.01	0.10 Average
Interfacial decohesion (B)	50–80 Short	240–310	< 0.01	0.10 Average
Rupture of fibers (C)	75–100 Long	> 300	< 0.01	0.05 Short
Peeling (D)	60–90 Short	260–295	> 0.02	> 0.15 Very long

Table 3.3. *Characteristics for classification descriptors*

Since the propagation of acoustic waves in materials depends on the elasticity modulus of materials or the direction of the reinforcements with respect to the direction of the load, for our materials, the values of the propagation speeds of their acoustic waves were 8,500 m/s, 4,500 m/s, 2,300 m/s and 1,320 m/s in agro-composites at [0°], [+/-45°], [+/-67.5°] and [90°], respectively. We could localize the acoustic bursts using two receivers and the propagation speed.

The distribution of AE bursts as a function of time during the tensile test is presented in Figure 3.20. The chronology of the appearance of different types of damage shows that the creation of micro-fissures in the matrix is the most notable damage mechanisms in traction tests on this type of material.

We should note that the creation of fissures in the matrix (A) corresponds to 70% of AE bursts, while decohesion (B) only corresponds to 30%. Three damage mechanisms have appeared since the start of tests; this damage then propagates along the length of the fibers in the form of intra-layer fissures and develops at the interface of folds, provoking the rupture of strata by peeling (type D). These results also demonstrate that the creation of fissures in the matrix remains preponderant and continues to develop after decohesion, peeling and fiber rupture. Type A and B signals are always in the majority with respect to type C (5%) and D (10%) signals.

For an "unknown" burst, the principle of the classification algorithm is to attribute it to the signal class from the library for which it has the largest number of nearest neighbors, comparing the signal to all the signals in the library. The number of neighbors is determined in advance by the user. Once this value has been defined, we only need to calculate the Euclidean distance for each AE signal with all of the signals in the library and search for its k-nearest neighbors. The signal will be labeled by evaluating the majority class among the nearest neighbors.

A supervised classification technique (k-nearest neighbors) is used to represent the different stages of damage and build the data library from signals collected during elementary tests:

– For tests at [0°], acoustic emissions are principally associated with the rupture of fibers.

– For tests at [+/-45°] and [67.5°], they are principally associated with peeling and fiber/matrix decohesion.

– For tests at [90°], only 10 signals were received at the coalescence of fiber/matrix decohesion and the rupture of the transverse matrix.

The order in which the four types of damage appear (A, B, C and D) in composites that are reinforced with hemp is presented in Figure 3.20. We can see that three damage mechanisms appear from the start of tests (t = 30 s) in the matrices, propagate from the cohesion interface and develop in the adjacent folds, thus provoking the rupture of the strata through peeling (type D). These also show that the appearance of fissures in the matrix remains preponderant and continues to develop after decohesion, peeling and fiber rupture. The signals from types A and B are always in the majority compared to the signals from types C and D.

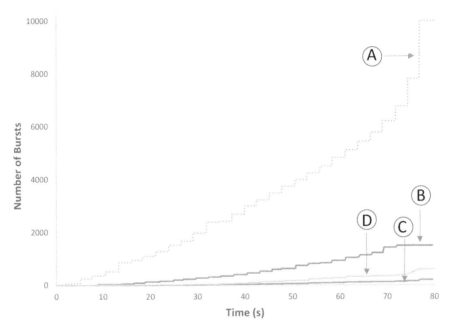

Figure 3.20. *Evolution in the number of AE signal bursts by damage class for composite materials*

3.5. Artificial neural networks for classifying damage

For damage identification and the classification of acoustic signatures, we must recognize signals of acoustic origin (Oja 1989; Johnson and

Gudmundson 2000; Dubuisson 2001; Zhang and Friedrich 2003; Muir et al. 2021). To solve this problem, we first decided to model the signal, before creating a classification. The time–frequency study of the signal demonstrated the usefulness of two different methods for modeling, one using average energy values per bandwidth and the other by auto-regressive modeling. Classification was performed using a network of multilayer neurons. It started by adapting these different parameters to the problem using chosen pairs of input–output points. Once the parameters of the network were stable, we then used it as a classifier. We will therefore succinctly describe how a multilayer network functions, and then present the signals we analyzed, as well as the results.

3.5.1. *Introduction*

The challenge in automatic classification is in developing computerized methods that learn to identify a new object for which we do not have any information a priori, using the knowledge we already have about a certain number of classes as a basis (a database of learning objects):

– An appropriate learning library database must be established.

– A precise rule for decision-making must be built to distinguish the different classes of learning objects.

– Classification of classes is represented by relevant descriptors.

However, relevant descriptors can be difficult to obtain or inefficient in constructing a decision-making rule. The nearest neighbor method is always costly in terms of calculation times. Another inconvenience is that its performance is potentially diminished when the scope of learning is small. To overcome these limitations and improve the precision of classification, this method must be replaced by decision-making rules that are more general, built using a learning set with a representative set for each class. *Deep learning* aims to improve the process of machine learning using *artificial intelligence (AI)*. The term "deep" refers to the many layers that the neural network accumulates while learning in order to improve its performance (Muir et al. 2021).

AI was born in 1950 to help humans automate certain everyday intellectual tasks by simply using a set of data and rules to detect and propose solutions based on the analysis of a predefined database. From the

1990s, machine learning appeared, which used statistical techniques to identify patterns within large quantities of data.

Computing developments in terms of analytic power over data gave way to the democratization of AI with the appearance of GPUs and optimized TPUs (NVIDIA Jetson Nano or Google Colab). In the field of AI, a network of artificial neurons is an organized set of interconnected neurons that help solve complex problems. The network of artificial neurons, or the neural network, is considered to be deep learning. Networks of artificial neurons are directly inspired by the way the nervous system works in analogy to the biological approach of neurons (Oja 1989; Schalkoff 1997; Adams and Hand 1999; Dubuisson 2001; Zhang and Friedrich 2003; Honeine and Richard 2007; Alpaydin 2009; Hand 2009; Bajaj and Bilas 2013; Fukunaga 2013).

Examples of applications are:

– industrial applications;

– recognition of area codes;

– control of processing parameters in industrial production;

– forecasting water consumption;

– decision-making software;

– weather forecasts.

The networks of artificial neurons are designed to reproduce some of their characteristics, such as the capacity to:

– learn;

– memorize information;

– analyze incomplete information.

A network of neurons depends on a large number of processors operating in parallel. Each artificial neuron receives input data in variable quantities; the origin of this input can be underlying neurons or receivers that make up the computer of which it is a part. A network of artificial neurons is made up of at least two layers, each of which contains several neurons. From one layer to the next, nodes are connected to one another. Input data, a weight, a threshold and output data are associated with each one.

The neuron illustrated in Figure 3.21 proceeds according to the following operations for each input:

1) weighting by multiplication of each input by connection weights;

2) summation of weighted input;

3) activation by passing the sum into a function (linear or nonlinear, or a threshold function);

4) the calculated value at the output of the neuron is sent to the subsequent neurons. If the output data of a node goes beyond the specified threshold, this neuron is activated and sends its data to the neurons of the next layer. And so on and so forth.

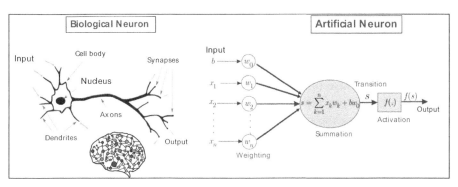

Figure 3.21. *Basic structure of a network of artificial neurons*

3.5.2. *Organization of neurons*

Neural networks (NN) are organized into three principal types of layers:

1) **the input layer**: this contains neurons whose values correspond to the multidimensional representation (a vector, matrix or structure with large dimensions) of the data we wish to analyze;

2) **the hidden layers**: these contain intermediary values calculated during learning in order to "capture" the variables that will make it possible to obtain the expected result from the input data;

3) **the output layer**: this contains neurons whose values are the desired result.

The principal objective of artificial neural networks is to minimize the distance between a prediction and the target value by aiming to adjust the weight little by little. With each data input that is submitted, the network must estimate the weight of the network. Artificial neural networks can be applied to multiple fields: image recognition, computer sight, vocal recognition, textual classification and transcription, automatic vehicles, autonomous robots, road and walkway surveillance, the generation of spoken language and even client-specific marketing (social networks, Youtube, etc.).

Deep learning designates the depth of layers in a network of artificial neurons. It is a particular type of automatic learning algorithm, like flowcharts or K-nearest neighbor, characterized by a large number of layers of neurons whose weighting coefficients are adjusted throughout the learning phase. We can distinguish three necessary phases in the "life" of a network (Figure 3.22):

1) **learning phase** (supervised or unsupervised): learn what relationship connects the vectors representing the network outputs to the explicative values representing the network's input signals, thus adapting the weight of connections to the present problem;

2) **validation phase**: verify whether the learned relationship in the preceding stage is relevant;

3) **test phase**: apply the network of neurons to a new input.

Three elements in the network can be modified:

1) the structure of the network (the number of connections entering each neuron, the number of hidden layers, etc.);

2) the type of network activation and transition functions;

3) the weight of the connections between neurons.

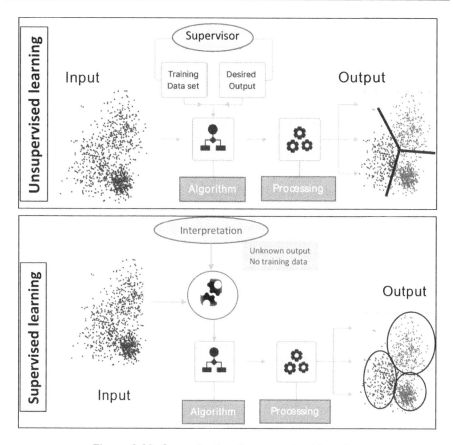

Figure 3.22. *Supervised and unsupervised learning*

3.5.3. *How the neural network functions*

Let there be a learning database $\left\{ X^{(a)}, Y^{(a)}, 1 \le a \le N \right\}$:

– a set of input signals $X^{(a)} = \left[X_1^{(a)}, X_2^{(a)}, \dots X_n^{(a)} \right]$;

– a set of corresponding output signals $Y^{(a)} = \left[Y_1^{(a)}, Y_2^{(a)}, \dots Y_m^{(a)} \right]$, the weighted sum of the signals;

– a set of output signals that were obtained empirically, $S^{(a)} = \left[S_1^{(a)}, S_2^{(a)}, \dots X_m^{(a)} \right]$;

– each connection that ties them to the next neuron is associated with a matrix of weights for the output layer ω;

– the simplest network is the monolayer one, called the perceptron. In the case of a **multi-class classification problem**, instead of using just one output unit, as many classes as are connected to all the input units must be used. Learning is an **iterative process** where after each observation, the weight of the connection must be adjusted to reduce the prediction error in the learning process. The algorithm of the gradient gives the direction of greatest variation for the error function so as to find the minimum of this function and move in the direction opposite to the gradient;

– we seek to minimize the overall error for each element in the learning database:

error:

$$E_k^{(a)} = S_k^{(a)} - Y_k^{(a)} \tag{3.15}$$

associated error:

$$E^{(a)} = \frac{1}{2} \sum_k \left(E_k^{(a)} \right)^2 \tag{3.16}$$

overall error:

$$E^{(a)} = \sum_a E^{(a)} \tag{3.17}$$

– we modify the network weights in the opposite direction using a gradient method:

$$\Delta \omega_{ij} = -\lambda \frac{\partial E^{(a)}}{\partial \omega_{ij}} \tag{3.18}$$

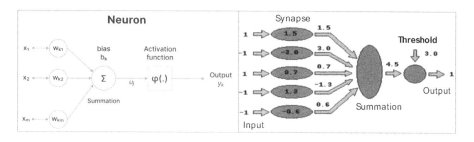

Figure 3.23. *General architecture of a NN*

The subscripts are related to the different layers: **i for the input layer, j for the intermediary layer** and **k for the output layer** (see Figure 3.23):

– the output is then obtained by applying the output neuron activation function in linear combination with the output from the intermediary layer:

- for the weights of the intermediary layers:

$$\Delta\omega_{kj} = \lambda \sum_{k} \left[\left(S_k^{(a)} - Y_k^{(a)} \right) \omega_{kj} f' \left(\sum_{j} \omega_{kj} f \left(\omega_{ji} X_i^{(a)} \right) \right) \right] f' \left(\sum_{i} \omega_{ji} X_i^{(a)} \right) X_i^{(a)} \quad [3.19]$$

- for the weights of the output layer:

$$\Delta\omega_{kj} = \lambda \left(S_k^{(a)} - Y_k^{(a)} \right) f \left(\sum_{i} \omega_{ji} X_i^{(a)} \right) f' \left(\sum_{j} \omega_{kj} f \left(\omega_{ji} X_i^{(a)} \right) \right) \quad [3.20]$$

– different types of neurons can be distinguished by the nature of the activation function:

1) linear: $f(x) = 1$,

2) threshold (Heaviside function):

$$f(x) = \begin{cases} 1 & \sum_{k=1}^{n} x_k \omega_k + b\omega_0 \geq 0 \\ 0 & \text{Otherwise} \end{cases}$$

3) sigmoid: $f(x) = 1/(1+e^{-x})$; provides an output between 0 and 1 to predict the probability of belonging to the positive class. The advantage of this function is that its derivative, an indispensable component of learning algorithms, can be easily obtained:

$$\frac{df(x)}{dx} = f(x)(1-f(x))$$

4) stochastic: $f(x) = 1$ with a probability of $1/(1+e^{-x/H})$, otherwise $f(x) = 0$.

The principal task of the multilayer neural network (or the *multilayer perceptron*, often called MLP) is to represent the entire analysis (modeling + estimation of parameters + filtering) so that the following blueprint becomes unnecessary for modeling:

Learning: minimize a cost function through gradient descent:

1) random initialization of parameters;

2) recursive calculation of the gradient using the retro-propagation formula;

3) sum over a mini-batch and update of parameters;

4) stop when the cost no longer decreases for disconnected data.

The gradient of the error function corresponds to the orientation and the slope at a given point:

– If the **norm of the gradient** is **high**, the slope is very steep.

– If the norm of the gradient is **low**, the slope is not very steep.

– If the gradient is **null**, it is flat (minimum).

– A gradient step that is too large or too small will make it impossible for the algorithm to converge towards an acceptable solution.

Finding the right gradient step is the entire challenge of machine learning and deep learning (see Figure 3.24):

– If λ is _large_, we are very far from ω_{ik}. If this point was already close to the optimal value, we risk surpassing our objective and ω_{ij} will be further than ω_{ik} from its optimal value. The algorithm risks **diverging**, that is, getting further from the optimal solution.

– If ω_{ik} is far from its optimal value and λ is _small_, the algorithm will take very long to **converge**.

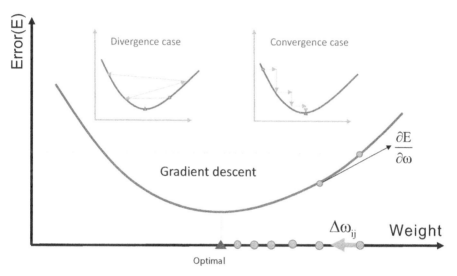

Figure 3.24. _Principle of mini-batch gradient descent_

3.5.4. *Results and analysis*

The collection of AE data (time to peak, amplitude, frequency, burst number, energy) and filtering in the general context of unsupervised AE machine learning, occurs as follows:

1) testing: 3,580 signals are collected;

2) extraction of characteristics: waveforms are represented in the realm of characteristics by extracting the relevant characteristics (2,000 signals);

3) artificial neural network algorithm: this partitions the waveforms into representative clusters corresponding to the damage mechanisms that account for the characteristics of agro-composites, as defined in Table 3.4;

4) labeling and error analysis: post-clustering analysis is performed to label the damage mechanism of clusters and evaluate the validity of the results.

Class	Amplitude A (dB)	Frequency (kHz)
Matrix crack (A)	35–65	90–180
Interfacial decohesion (B)	50–80	240–310
Rupture of fibers (C)	75–100	> 300
Peeling (D)	60–90	260–295

Table 3.4. *Amplitude and frequency: classification parameters for damage*

To establish the learning library database for hemp–PP agro-composites, the neural network method was used to classify the AE signals obtained from the tests on the tests tubes as models for these composites. The classification of 3,580 signals from the tests comes from the comparison of the signals from the test to those in the library.

We applied the NN method on each of the signals in the library in order to isolate the signals that are not assigned to their original class. This allows us to create a library based only on the signals that are classified according to their original class. In this study, for agro-composites with 30% hemp fibers, 1,537 signals were thus removed as they corresponded to minority sources. For example, for a test at $0°$, we cannot exclude decohesion and peeling. In this nonlinear example, we consider a data set where each input vector is associated with one of the four classes: the creation of fissures in the matrix is **class 1**, interfacial decohesion is **class 2**, the creation of fissures in the matrix–fiber interface is **class 3** and damage to the fibers is **class 4**.

Figures 3.25 and 3.26 illustrate the distribution of received signals in amplitude and frequency as a function of time, which shows that there are three distinct populations. The first with an amplitude varying between 35 and 55 dB corresponds to type A matrix damage, the one between 55 and 70 dB corresponds to type B matrix–fiber debonding and the one above 70 dB corresponds to type C fiber damage. Figure 3.27 shows the results of this classification through projection (amplitude, frequency). The results indicate that among the different AE descriptors, frequency is the most distinguishable parameter; consequently, it can be used as an efficient AE parameter for the classification of damage in composites. As mentioned above, the highest frequency and the lowest frequency in the ranges are representative of the rupture of fibers and the rupture of matrix fissures, respectively.

The spread of matrix–fiber interfacial decohesion frequencies is situated between the rupture of the fiber and the matrix. However, the overlap between the two classes, 2 and 3, of interface damage is more pronounced. This overlap is due to the presence of similar AE signals between the types of damage to the interfaces of composite materials. The AE signals of all the types of damage were classified with a precision of 91% (see Table 3.5). In conclusion, we see that for the database of hemp fiber composite materials, the algorithm of neural networks makes it possible to classify AE signals for the damage mechanisms in the matrix, fibers and interfaces.

	% signals [0°]	% signals [+/-45°]	% signals [+/-67.5°]	% signals [90°]
Class A	4.00%	10.18%	30.15%	18.68%
Class B	7.31%	**70.94%**	**66.35%**	37.42%
Class C	**78.3%**	3.30%	1.50%	0.55%
Class D	11.69%	21.67%	16.55%	**45.88%**

Table 3.5. *Percentage of signals for each class in agro-composite tests*

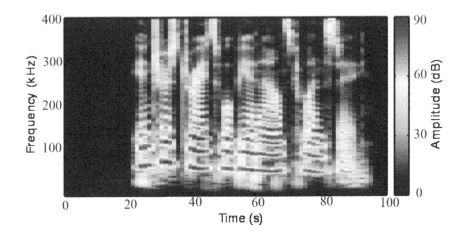

Figure 3.25. *Time–frequency–amplitude map of AE signals in agro-composites*

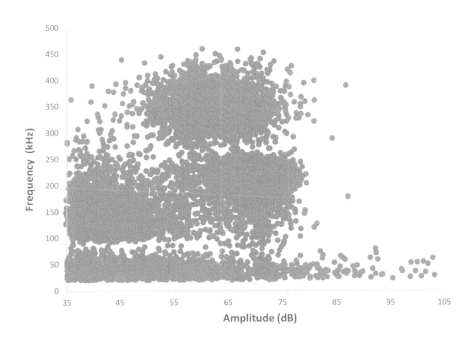

Figure 3.26. *Frequency–amplitude before damage classification for agro-composites*

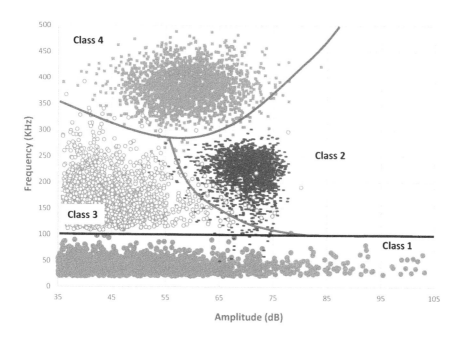

Figure 3.27. *Projection of damage classes for agro-composites*

3.6. Conclusion

Different damage mechanisms occur almost simultaneously during composite loading. AE monitoring is one of the most suitable methods for detecting damage occurrence and its evolution in real time during the loading of fiber-reinforced polymers. High sensitivity of AE monitoring offers detection of different damage sources, such as transverse matrix cracking as the first occurring damage mechanism in composite laminates, delamination, fiber failure, debonding between the matrix and fibers, and fiber pull-out.

Assigning a specific set of AE signal features to a particular damage mechanism is a scientific challenge. Various machine learning techniques are often used in combination with feature selection to extract frequency descriptors. For the automatic extraction of the intrinsic characteristics of signals, deep learning methods are becoming increasingly popular. While feature values are traditionally obtained by manual construction and require a certain professional knowledge, deep learning is used to automatically extract deep features, avoiding the loss of signals in time and frequency domains during the manual feature extraction.

Three actions related to current classification methods for AE signals have been underscored:

– identification of classes required by unsupervised classifications;

– validation of descriptors used to characterize AE signals;

– establishment of a learning database of AE signals through supervised methods.

This study was conducted with a set of laboratory experiments that examined samples of bio-composites and hemp fiber–PP composites. It allows us to distinguish the damage to each of the components, namely the matrix, the hemp fiber–matrix interfaces, peeling of folds and rupture of plant fibers.

The implementation of the proposed method was based on the temporal and frequential characteristics of AE signals through unsupervised statistical approaches (K-means). In order to further improve the performance of classification approaches, we ended by proposing that the input data should be enriched with information about the context of each emission using supervised data analysis techniques (neural networks).

The results are indicative of the potential of the proposed methodology for the characterization of loaded composite materials. However, further research is needed to examine a wider range of bio-composite structures in different loading and environmental conditions.

3.7. References

Adams, N.M. and Hand, D.J. (1999). Comparing classifiers when the misallocation costs are uncertain. *Pattern Recognition*, 32(7), 1139–1147.

Alpaydin, E. (2009). *Introduction to Machine Learning*. MIT Press, Cambridge.

Antony, S., Cherouat, A., Montay, G. (2018). Experimental, analytical and numerical analysis to investigate the tensile behaviour of hemp fibre yarns. *Composite Structures*, 202, 482–490.

Antony, S., Cherouat, A., Montay, G. (2019). Multiscale analysis to investigate the mechanical and forming behaviour of hemp fibre woven fabrics/polypropylene composite. *ICCM 2019 – 22nd International Conference on Composite Materials*, August, Melbourne, 318–327. hal-02473963.

Ashori, A. (2008). Wood–plastic composites as promising green-composites for automotive industries! *Bioresource Technology*, 99, 4661–4667.

Awerbuch, J., Leone, F., Ozevin, D., Tan, T.-M. (2016). On the applicability of acoustic emission to identify modes of damage in full-scale composite fuselage structures. *Journal of Composite Materials*, 50(4), 447–469.

Aymond, R. and Almer, J.P. (2012). Composite processes in aeronautics. *Wiley Encyclopedia of Composites*, 1–24.

Bajaj, V. and Pachori, R.B. (2013). Automatic classification of sleep stages based on the time-frequency image of EEG signals. *Computer Methods and Programs in Biomedicine*, 112(3), 320–328.

Baley, C. and Lamy, B. (2000). Propriétés mécaniques des fibres de lin utilisées comme renforts de matériaux composites. *Revue des composites et des matériaux avancés*, 10, 7–24.

Baley, C., Grohens, Y., Pillin, I. (2004). État de l'art sur les matériaux composites biodégradables. *Revue des composites et des matériaux avancés*, 2, 135–166.

Benzeggagh, M.L., Barre, S., Echalier, B., Jacquemet, R. (1992). Etude de l'endommagement de matériaux composites à fibres courtes et à matrice thermoplastique. *AMAC Journées nationales composites*, Paris, 8, 703–714.

Berthelot, J.-M. (2012). *Matériaux composites : comportement mécanique et analyse des structures*, 5th edition. Lavoisier Éditions Tec & Doc, Paris.

Berthelot, J.-M. and Rhazi, J. (1990). Acoustic emission in carbon fibre composites. *Composites Science and Technology*, 37(4), 411–428.

Ceysson, O., Salvia, M., Vincent, L. (1996). Damage mechanisms characterization of carbon fibre/epoxy composite laminates by both electrical resistance measurements and acoustic emission analysis. *Scripta Materialia*, 34(8), 1273–1280.

Chaki, S., Harizi, W., Krawczak, P., Bourse, G., Ourak, M. (2016). Structural health of polymer composite: Nondestructive diagnosis using a hybrid NDT approach. *JEC Composites Magazine*, 53(107), 62–65.

Chen, O., Karandikar, P., Takeda, N., Kishi, T. (1992). Acoustic emission characterization of a glass-matrix composite. *Nondestructive Testing and Evaluation*, 8–9, 869–878.

Dieulesaint, E. and Royer, D. (1974). *Ondes elastiques dans les solides : application au traitement du signal*. Masson, Paris.

Dubuisson, B. (2001). *Diagnostic, intelligence artificielle et reconnaissance des formes*. Hermés Science, Paris.

Farrar, C. and Worden, K. (2013). *Structural Health Monitoring: A Machine Learning Perspective*. John Wiley & Sons Ltd, Chichester.

Farzad, P. and Ramin, K. (2021). Damage classification of sandwich composites using acoustic emission. *npj Computational Materials*, 7, 95. doi: 10.1038/s41524-021-00565-x.

Fukunaga, K. (2013). *Introduction to Statistical Pattern Recognition*. Elsevier, Oxford.

Godin, N., R'Mili, M., Reynaud, P., Lamon, J., Fantozzi, G. (2011). Emission acoustique et endommagement des composites : intérêts et limites des techniques de reconnaissance de forme. *JNC 17*, Poitiers.

Gorman, M.R. and Prosser, W.H. (1991). AE source orientation by plate wave analysis. *Journal of Acoustic Emission*, 9, 283–288.

Hamdi, S.E., Le Duff, A., Simon, L., Plantier, G., Sourice, A., Feuilloy, M. (2013). Acoustic emission pattern recognition approach based on Hilbert–Huang transform for structural health monitoring in polymer-composite materials. *Applied Acoustics*, 74, 746–757.

Hand, D.J. (2009). Measuring classifier performance: A coherent alternative to the area under the ROC curve. *Machine Learning*, 77(1), 103–123.

Harizi, W. (2012). Caractérisation de l'endommagement des composites à matrice polymère par une approche multi-technique non destructive. PhD Thesis, Mines Douai and Univ., Valenciennes.

Honeine, P. and Richard, C. (2007). Signal-dependent time frequency representations for classification using a radially Gaussian kernel and the alignment criterion. *IEEE/SP14th Workshop on Statistical Signal Processing, SSP'07*, 735–739.

Huguet, S. (2002). Application de classificateurs aux données d'émission acoustique : identification de la signature acoustique des mécanismes d'endommagement dans les composites à matrice polymère. PhD Thesis, INSA Lyon, Villeurbanne.

Huguet, S., Godin, N., Gaertner, R., Salmon, L., Villard, D. (2002). Use of acoustic emission to identify damage modes in glass fibre reinforced polyester. *Composites Science and Technology*, 62, 1433–1444.

Ilczyszyn, F., Cherouat, A., Montay, G. (2012). Caractérisation des propriétés mécaniques de fibres de chanvre. *Matériaux et Techniques*, 100(5), 451–457.

Johnson, M. and Gudmundson, P. (2000). Broad-band transient recording and characterization of acoustic emission events in composite laminates. *Composites Science and Technology*, 60, 2803–2818.

Keprt, J. and Benes, P. (2008). A comparison of AE sensor calibration methods. *Journal of Acoustic Emission*, 26, 60–70.

Khamedi, R., Abdi, S., Ghorbani, A., Ghiami, A., Erden, S. (2020). Damage characterization of carbon/epoxy composites using acoustic emission signals wavelet analysis. *Composite Interfaces*, 27(1), 111–124. doi: 10.1080/09276440. 2019.1601939.

Kim, S.-T. and Lee, Y. (1997). Characteristics of damage and fracture process of carbon fiber reinforced plastic under loading-unloading test by using AE method. *Materials Science and Engineering*, 234–236, 322–326.

Kotsikos, G., Evans, J.T., Gibson, A., Hale, J. (1999). Use of acoustic emission to characterize corrosion fatigue damage accumulation in glass fiber reinforced polyester laminates. *Polymer Composites*, 20(5), 689–696.

La Mantia, F.P. and Morreale, M. (2011). Green composites: A brief review. *Composites Part A: Applied Science and Manufacturing*, 42(6), 579–588.

Liao, T.W. (2005). Clustering of time series data – A survey. *Pattern Recognition*, 38(11), 1857–1874.

Marec, A., Thomas, J.-H., El Guerjouma, R. (2008). Contrôle de santé des matériaux hétérogènes par émission acoustique et acoustique non linéaire : discrimination des mécanismes d'endommagement et estimation de la durée de vie restante. PhD Thesis, Laboratoire d'Acoustique and Le Mans University, Le Mans.

May, Z., Alam, M.K., Mahmud, M.S., Rahman, N.A.A. (2020). Unsupervised bivariate data clustering for damage assessment of carbon fiber composite laminates. *PLoS ONE*, 15(11), e0242022. doi: 10.1371/journal.pone.0242022.

Mohamad, I., Milad, H., Mehdi, A. (2014). Technique and k-means genetic algorithm. *Journal of Nondestructive Evaluation*, 33(4). doi: 10.1007/s10921-014-0243-y.

Muir, C., Swaminathan, B., Almansour, A.S., Sevener, K., Smith, C., Presby, M., Kiser, J.D., Pollock, T.M., Daly, S. (2021). Damage mechanism identification in composites via machine learning and acoustic emission. *npj Computational Materials*, 7, 95. doi: 10.1038/s41524-021-00565-x.

Ohtsu, M. and Ono, K. (1988). AE source location and orientation determination of tensile cracks from surface observation. *NDT International*, 21, 143–150.

Oja, E. (1989). Neural networks, principal components, and subspaces. *International Journal of Neural Systems*, 1, 61–68.

Ono, K. (1997). Acoustic emission. In *Encyclopedia of Acoustics*, Crocker, M.J. (ed.). John Wiley & Sons Inc, New York.

Pappas, Y.Z. and Kostopoulos, V. (2001). Toughness characterization and acoustic emission monitoring of a 2D carbon/carbon composite. *Engineering Fracture Mechanics*, 68(14), 1557–1573.

Pickering, K. (2008). *Properties and Performance of Natural Fibre Composites*. Elsevier, Amsterdam.

Proust, A., Wargnier, H., Harry, R., Lorriot, T., Lenain, J.C. (2006). Détermination d'un critère d'amorçage du délaminage au sein d'un matériau composite à l'aide de la technique d'émission acoustique. *Matériaux*, 13–17.

Roundi, W., El Mahi, A., El Gharad, A., Rebiere, J.-L. (2018). Acoustic emission monitoring of damage progression in glass/epoxy composites during static and fatigue tensile tests. *Applied Acoustics*, 132, 124–134.

Saeedifar, M. and Zarouchas, D. (2020). Damage characterization of laminated composites using acoustic emission: A review. *Composites Part B: Engineering*, 108039. doi: 10.1016/j.compositesb.2020.108039.

Schalkoff, R.J. (1997). *Artificial Neural Networks*. McGraw-Hill, New York.

Scida, D., Aboura, Z., Benzeggagh, M.L. (2002). The effect of ageing on the damage events in woven-fibre composite materials under different loading conditions. *Composites Science and Technology*, 62, 551–557.

Scruby, C. (1987). An introduction to acoustic emission. *Journal of Physics E: Scientific Instruments*, 20, 946.

Shiino, M.Y., Faria, M.C.M., Botelho, E.C., de Oliveira, P.C. (2012). Assessment of cumulative damage by using ultrasonic C-scan on carbon fiber/epoxy composites under thermal cycling. *Materials Research*, 15(4), 495–499.

Šofer, M., Cienciala, J., Fusek, M., Pavlicek, P., Moravec, R. (2021). Damage analysis of composite CFRP tubes using acoustic emission monitoring and pattern recognition approach. *Materials*, 14, 786. doi: 10.3390/ma14040786.

Tabrizi, I.E., Kefal, A., Zanjani, J.S.M., Akalin, C., Yildiz, M. (2019). Experimental and numerical investigation on fracture behavior of glass/carbon fiber hybrid composites using acoustic emission method and refined zigzag theory. *Composite Structures*, 223, 110971.

Tat, H., Wu, J., Pike, M., Schaefer, J., Pauca, V.P., Li, R. (2017). Machine learning for acoustic emission signatures in composite laminates. *32nd Annual American Society for Composites Technical Conference*, 2, 1235–1251.

Touya, R. (1979). Contrôle non destructif par émission acoustique. Report, Directions des Recherches Etudes et Techniques.

Tuloup, C., Harizi, W., Aboura, Z., Meyer, Y. (2019). Structural health monitoring of smart polymer-matrix composite during cyclic loading using an in-situ piezoelectric sensor. *ICCM22: 22nd International Conference on Composite Materials*, Melbourne. hal-02963212.

Wambua, P., Ivens, J., Verpoest, I. (2003). Natural fibres: Can they replace glass in fibre reinforced plastics? *Composites Science and Technology*, 63, 1259–1264.

Yuyama, S. (1986). Fundamental aspects of acoustic emission applications to the problems caused by corrosion. In *Corrosion Monitoring in Industrial Plants Using Non-Destructive Testing and Electrochemical Methods*, Moran, G.C. (ed.). ASTM International, West Conshohocken.

Zhang, Z. and Friedrich, K. (2003). Artificial neural networks applied to polymer composites: A review. *Composites Science and Technology*, 63, 2029–2044.

Zhou, W., Zhao, W.-Z., Zhang, Y.-N., Ding, Z.-J. (2018). Cluster analysis of acoustic emission signals and deformation measurement for delaminated glass fiber epoxy composites. *Composite Structures*, 195, 349–358.

4

Intelligent Control for Attenuating Vertical Vibrations in Vehicles

In the automotive field, advances in control tools have considerably improved performance in terms of both comfort and security. These developments depend on behavioral models of the mechanical groupings that make up the vehicle, created using approaches based on multi-body dynamics and more or less complex models that occasionally present very marked nonlinearities. Moreover, the possibilities for activation multiply with the diversification of vehicular architecture (traction chain, motorization, suspension, steering axles, etc.). Now, the objective thus consists of applying and stretching developments proposed in the context of model-free control (MFC) to dynamic behavioral models of mechanical assemblages, that is, vehicles. Several designers have recently come around to adapting these developments to applications for which there is a rather strong coupling between the degrees-of-freedom taken and the actuators.

4.1. Introduction

Feedback control of complex systems with nonlinearities remains difficult despite the progress of the last decades. On the one hand, the structure of the controller is still sensitive to unknown variations and requires the designer's intervention in order to recalibrate the parameters of the controller. On the other hand, measurements are generally noisy or unattainable. Consequently, the use of very expensive instruments such as laser receivers or filters is

Chapter written by Maroua HADDAR, Fakher CHAARI and Mohamed HADDAR.

strongly recommended. Indeed, we should not forget the application of control strategies based on models of real systems that present several well-known difficulties such as complexity, uncertainty, nonlinearity, variability and temporal dynamics. Most of the traditional control techniques refer to factory models that are simplified and come from hypotheses that differ from one designer to the next. For example, a law of nonlinear control must be derived from a model using techniques such as flatness-based control (Fliess et al. 1995), linearizing the system on the basis of its operational value or using a chain of systems (Ren and Beard 2008). Obtaining a model, however, can be either too expensive due to the engineering time and materials required, or simply impossible, as with the braking system of vehicles.

In the automotive industry, the suspension system plays a role in ensuring the stability of the vehicle while it moves along on roads. On numerous modern vehicles, an active suspension system has been proposed as a replacement for the conventional passive suspension system. The performance of active suspension system controllers depends on the method of control. In fact, technological advancements in recent years have attempted to facilitate driving by developing active security or assisted-driving systems (Haddar et al. 2019a).

These systems, called intelligent systems, aim to facilitate vehicular control and optimize the tasks involved in driving by providing drivers with advanced warnings of dangerous situations. The systems have the capacity to identify different critical situations when online by referring to well-determined risk functions. Let us take the example of road surfaces, which are the source of the majority of perturbances, and which deteriorate the controller attached to the suspension actuators. These unknown stimuli have grabbed the attention of automotive producers and the government over the last couple of decades. There are several methods that use visual inspections to determine the depth of the road using video cameras or depth sensors and send the collected information to the electronic control unit (Ni et al. 2020).

The efficiency of an active suspension system is entirely dependent on the method of control. Recently, several types of research projects have been published on the question of active suspension system control. Among them, several methods for linear control have been used. A dual loop PID controller method was also proposed (Shafie et al. 2015). This technique is composed of one internal input loop and one external output loop. The external input loop gives the instruction signal for the internal input loop. Frequent use of PID regulators is generally justified by the fact that good modeling is unnecessary. However, measuring the advantages of PID is difficult and takes much time. This conventional linear controller is not able to respond to the demands of vehicular stability. Consequently, a model-free control approach was proposed (Haddar et al. 2021) and successfully applied to the active control of the vehicle's dynamic suspension (Menhour et al. 2015).

The notion of "model-free control" was developed by Michel Fliess and Cédric Join. It is based on the use of algebraic differentiators and has turned out to be a practical approach for solving difficult problems. The use of this type of differentiator in feedback control began with the linear systems in the work of Fliess and Sira-Ramírez (2003) and was extended to nonlinear systems in their subsequent work (Fliess and Sira-Ramírez 2004). Several successful applications using this approach can be found in the literature. Model-free control of the joints in humanoid robots was discussed in Villagra and Balaguer (2009). In 2012, the approach was applied to the control of an inverted inertia wheel (Andary et al. 2012). For automotive systems, in 2014, Fergani et al. (2014) used the estimated derivatives of the components of a flat offramp in combination with an observer to reconstruct the surface of a road taken by a utility vehicle equipped with four magnetorheological shock absorbers. These estimations were then used to develop a control strategy.

Control engineers can become disconcerted by the automatic learning strategies that study how historical data can be used to improve the future operations of dynamic systems. They posit that this is precisely the significance of control theory. Fliess and Join (2020) classified MFC as a new technique for machine learning (ML). According to the works cited, MFC is easy to implement and can be replaced in control engineering by ML through artificial neuron networks and/or learning through reinforcement.

4.2. Limits of passive and semi-active control strategies

4.2.1. *Limits of passive suspension*

The first techniques to be established were passive suspension, generally used in numerous fields. Passive controllers are used due to their relative simplicity to operate and the capacity of their components to produce potential contributions. Moreover, they are reputed for their unconditionally stable nature and their independence from any outside source of energy.

Passive approaches are efficient in the bandwidths of medium and high frequencies. They function with little success at low frequencies, however. The ability to convert passive control is determined by the properties of the elements in the process such as the shock absorption coefficient and the rigidity coefficient. The characteristics of shock absorbers and springs can be selected but they are fixed. Because of this, these approaches are not sufficiently efficient for cases of changes in the type of road surface, which require adaptability for each point of operation. This lack of adaptability can be compensated for by designs based on a great amount of experience in industry. For these reasons, in order to face the conflicting demands of the dynamics of vehicular suspension, several studies have been conducted using other strategies such as active controllers and semi-active controllers.

4.2.2. *Limits of semi-active suspension*

In order to mitigate the disadvantages of passive control, other techniques such as semi-active control were developed. Currently, these improvements are being developed as new intelligent materials with modified properties are arriving on the market and are taken into account. Consequently, depending on the operational conditions, the characteristics of the mechanical organs (shock absorption, mass distribution, stiffness) are modified: for example, the characteristics of shock absorbers have been turned on their head. Moreover, there is no need for actuation energy. There are therefore no actuators in the sense of active force; only the shock absorber is necessary for modifying the overall characteristics. Usually, semi-active suspension can be actively controlled by a control unit. The modification of the shock absorption coefficient may be continuous or discontinuous.

Most suspension control strategies were developed in one of two ways, as in the Skyhook shock absorption proposed by Karnopp (1995) or the Groundhook shock absorption proposed by Valášek et al. (1997). We can see that semi-active suspension works with closed-loop feedback control and is used as a reference model (Zhang et al. 2013).

4.3. Model-free control (MFC)

This active control is mostly used in industrial applications where the levels of vibration are high. Contrary to passive control and semi-active control, active control requires external operation or operational energy. This energy is related to the power of actuators and amplifiers. Thus, the control power and its efficiency are only limited by the choice of components and control algorithms.

The performance of an MFC algorithm must be characterized by the following three indicators.

– Robustness: the control technique must perform in the face of a large class of perturbances, namely measurement noise, unknown input, uncertainty over parameters, etc.

– Improvements on calculation time: obviously, the control strategy must have as short a calculation time as possible in order to provide quick estimates in real time and online for processes with fast dynamics.

– Easy conception and implementation: the controller and observer must have as little complexity as possible. The engineer must find the appropriate way of applying algorithms (i.e. scaling parameters must have a physical meaning).

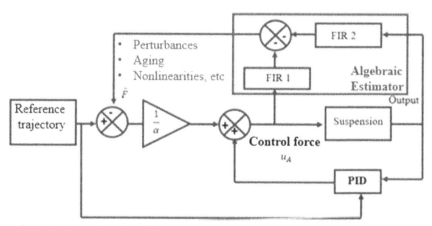

*FIR: Algebraic Low Pass Filter
*PID: A proportional–integral–derivative controller

Figure 4.1. *Model-free control block diagram. For a color version of this figure, see www.iste.co.uk/elhami/uncertainty.zip*

The power of the intelligent controller lies in the online perturbance estimates that do not require state observation. The method is easily implemented and avoids the difficult task of calibrating coefficients. Only a knowledge of the input–output behavior of the process is required to automatically update the parameters of the regulator (see Figure 4.1). The theoretical context of this technique will be explained subsequently.

4.3.1. *Ultra-local model*

The principle of the model-free control paradigm introduced by Fliess and Join (2013) is to substitute a nonlinear complex model by an ultra-local model that only applies over a very short period of time, L:

$$y^{(n)}(t) = F(t) + \alpha u(t) \tag{4.1}$$

where u is the control force, y is the observed or measured output, n is the system order, F includes both the unexpected perturbances and the non-modeled dynamic and α is a parameter chosen by the designer. In practice, the order n is always 1 or 2.

4.3.2. *Online estimates of modeled and non-modeled perturbances*

The integration of the model-free control approach imposes certain implementation requirements that must be carefully followed. We begin with the obligation to use a high frequency with respect to the frequency of the system we wish to control, $f_s = \dfrac{1}{T_s}$ (where T_s is the time step for sample measures). In fact, at each time step $t_k = k.T_s$, the estimation process of the dynamic, F, is calculated from the difference between the previous command u applied to the system and the observed output y. At each instant t, the estimate of F is calculated to give rise to $\hat{F}$, which is meant to be constant over the interval $[t, t+L]$ (see Polack 2018). Simply speaking, we can express the estimate of $\hat{F}_k$ by:

$$\hat{F}_k = y_k^{(n)} - \alpha.u_{k-1} \tag{4.2}$$

However, the measurements made by proprioceptive sensors installed in the vehicle are generally noisy. Consequently, the use of a filter is strongly recommended. Moreover, the estimate of derivatives from measured signals is obligatory, given the lack of sensory materials. The estimation methods for $\hat{F}$ are different from one designer to the next. We will now revisit the different versions of model-free control.

4.3.2.1. Algebraic estimator, type 1

In the first version, $\hat{F}$ was built on the difference between an estimate of the derivative of the order n from measured signals and the previous command to be applied to the controlled system:

$$\left[\hat{F}(t) \right] = \left[\hat{y}^{(n)}(t) \right] - \alpha u_A(t-1) \qquad [4.3]$$

The quality of online numerical derivatives of noisy signals plays an important role in different applications in diverse fields. Sidhom (2011) studied the algebraic derivators and developed sliding mode derivators to correct the limits of the first versions of algebraic differentiators. This study demonstrates that there are three versions of algebraic differentiators. None of the versions present convergence issues given that they have an algebraic form. Moreover, these algorithms are independent of the type of measurement noise. The complexity comes, however, in the implementation stage since there is a high number of calibration parameters that come into play. These non-asymptotic, numerically approximative approaches for derivatives of noisy signals of a superior order are adapted to systems operating in real time. The different approaches to derivation are summarized and evaluated in terms of their historical evolution by Othmane et al. (2022).

4.3.2.2. Algebraic estimator, type 2

The unknown function F can be approximated by a constant function in parts. According to Fliess et al. (2013), F is a linearly identifiable constant. By applying the notions of operational calculations mentioned in Yosida (2012) in equation [4.1], with $n = 2$, we get:

$$\hat{F}_k = -\frac{5!}{2L^5} \int_0^L \left[\left(-L^2 + 6L\tau - 6\tau^2 \right) y(\tau) + \frac{\alpha}{2} \tau^2 \left(L - \tau \right)^2 u(\tau) \right] d\tau \qquad [4.4]$$

If n = 1, we get:

$$\hat{F}_k = -\frac{6}{L^3}\int_0^L \left[(L-2\tau)y(\tau)+\alpha\tau(L-\tau)u(\tau)\right]d\tau \qquad [4.5]$$

4.3.2.3. Averaging method, type 3

This approach could be called the "averaging" method. The formula is published; however, it is not really used. Let us suppose that F is almost constant over a short period (Fliess et al. 2011):

$$\hat{F}_k \approx \frac{1}{\gamma}\int_{L-\gamma}^L \dot{y}^* d\tau - \frac{\alpha}{\gamma}\int_{L-\gamma}^L u_A d\tau - \frac{K_P}{\gamma}\int_{L-\gamma}^L e\, d\tau \qquad [4.6]$$

4.3.3. Intelligent controller

The difference between the formulation of conventional PID controllers and the intelligent controllers presented here is the presence of the term $\hat{F}_k$. The online estimate is used to cancel endogenous and exogenous perturbances in the system. If the order of differentiation n in the ultra-local model in equation [4.1] is taken to be equal to 2, the feedback loop is closed by the intelligent i-PID proportional-integral derivative controller, as indicated below:

$$u(t) = -\frac{\hat{F}}{\alpha} + \frac{\ddot{y}^* - \left(K_P e + K_I \int e + K_D \dot{e}\right)}{\alpha} \qquad [4.7]$$

where u is the force of the actuator, e is the monitoring error, $\ddot{y}^*$ is the reference to follow, and K_P, K_I and K_D are the gains of the PID controller (proportional gain, integral gain and derivative gain, respectively).

The intelligent controller in equation [4.2] contains two parts: a new term to cancel the perturbance and the classic controller term. Consequently, the i-PID is potentially applicable to unmodeled problems when compared with the classic PID controller.

If $K_I = 0$, we get an i-PD. Moreover, if $n = 1$, we call the controller intelligent i-PI or i-P.

We call them "intelligent" because of the ability of these regulators to bring asymptotic stability to the system in the absence of any estimate errors, even in the transitional phases where the reference signal is not constant, contrary to the classical PID regulators.

4.4. Application to the control of vertical vehicular vibrations

4.4.1. *Performance demands*

The quality of a vehicle's dynamic performance can be measured through set objectives. The following factors must be quantified in order to allow the engineer to compare different suspension designs and control strategies for active and passive suspensions. Consequently, the performance demands on vehicular suspension include the following aspects:

Driving comfort: improvement in driving comfort is one of the principal tasks of active suspension. Driving comfort is therefore an important measure of performance for the design of a vehicle, which requires the design of a controller capable of stabilizing the vertical movement of the car body and isolating the vibrations transferred to passengers.

Grip on the road: an important demand for ensuring safety when driving is to make sure the contact between the tire and the road is good. When the transfer of longitudinal force is activated, the driver must be able to control the vehicle through directional, accelerating and braking inputs. Consequently, in order to ensure safety and to improve the quality of how the car drives, the dynamic load on the wheels must be limited and not surpass the static load on them.

4.4.2. *Complete vehicular model*

Figure 4.2 shows a diagram of the integral suspension system with seven degrees of freedom that we are considering. The following mathematical model describes the movement of the suspended mass (vertical, oscillatory and rotational) and the dynamic equations of the bouncing movement of the tires:

$$
\begin{cases}
m_c \ddot{z}_c = -\sum_{i=1}^{4}(F_{ki} + F_{di} + f_i) \\[2mm]
I_y \ddot{\theta}_c = a_1 \sum_{i=1}^{2}(F_{ki} + F_{di} + f_i) - a_2 \sum_{i=3}^{4}(F_{ki} + F_{di} + f_i) \\[2mm]
I_x \ddot{\varphi}_c = b_1 \sum_{i=1,3}(F_{ki} + F_{di} + f_i) - b_2 \sum_{i=2,4}(F_{ki} + F_{di} + f_i) \\[2mm]
m_{u1} \ddot{z}_{u1} = -F_{ku1} - F_{cu1} + F_{k1} - F_{d1} + f \\[1mm]
m_{u2} \ddot{z}_{u2} = -F_{ku2} - F_{cu2} + F_{k2} - F_{d1} + f_2 \\[1mm]
m_{u3} \ddot{z}_{u3} = -F_{ku3} - F_{cu3} + F_{k3} - F_{d3} + f_3 \\[1mm]
m_{u4} \ddot{z}_{u4} = -F_{ku4} - F_{cu4} + F_{k4} - F_{d4} + f_4
\end{cases}
\qquad [4.8]
$$

where z_c, θ_c and φ_c are the vertical displacement, the oscillatory angle and the rotational angle of the vehicle's body, respectively. F_{di} is the shock absorption force at each corner. F_{ki} is the elastic force of each suspension. The force of the tire is represented by F_{kui}. The shock absorption force at each corner of the vehicle is given by F_{ci} and will be negligible in this study. The expression of these forces with nonlinear models is given by:

$$
\begin{cases}
F_{ki} = k_i (z_i - z_{ui}) \\[1mm]
F_{di} = d_i (\dot{z}_i - \dot{z}_{ui}) \\[1mm]
F_{kui} = k_{ui}(z_{ui} - z_{ri})
\end{cases}
\qquad [4.9]
$$

Nonlinearities are always present in practice; they come from problems in the springs or leaks in the shock absorbers. Consequently, the expression of force changes and can be modeled as follows:

$$
\begin{cases}
F_{ki} = k_i (z_i - z_{ui}) + k_{nli}(z_i - z_{ti})^3 \\[1mm]
F_{di} = d_i (\dot{z}_i - \dot{z}_{ui}) + d_{nli}\sqrt{(\dot{z}_i - \dot{z}_{ui})}\, sign(\dot{z}_i - \dot{z}_{ui})
\end{cases}
\qquad [4.10]
$$

where a_1 is the spacing between the front axle and the center of gravity, G, of the vehicle, a_2 is the distance between the back axle and the center of gravity, G, b_1 is the spacing between the front left suspension and the rotational axis, and b_2 is the distance between the front right suspension and the rotational axis. The values and notations of the complete parameters for a car are detailed in Bouazara et al. (2006).

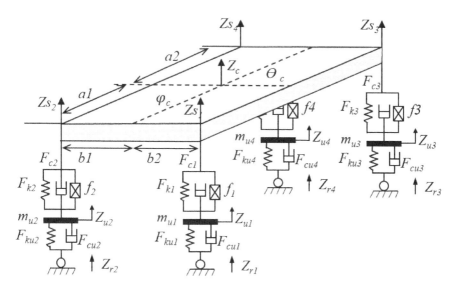

Figure 4.2. *Integral suspension system with seven degrees of freedom. For a color version of this figure, see www.iste.co.uk/elhami/uncertainty.zip*

4.4.3. *Ultra-local models*

The configuration of the transformation of input decoupling is given by three systems with one unique input and one unique output that have replaced the MIMO system (Multiple-Input Multiple-Output) given by equation [4.1] as follows:

$$
\begin{cases}
\ddot{z}_c = \zeta_{zc} + \dfrac{d_{11}f_1 + d_{12}f_2 + d_{13}f_3 + d_{14}f_4}{m_c} \\[4mm]
\ddot{\theta}_c = \zeta_{\theta c} + \dfrac{d_{21}f_1 + d_{22}f_2 + d_{23}f_3 + d_{24}f_4}{I_y} \\[4mm]
\ddot{\varphi}_c = \zeta_{\varphi c} + \dfrac{d_{31}f_1 + d_{32}f_2 + d_{33}f_3 + d_{34}f_4}{I_x}
\end{cases}
\qquad [4.11]
$$

The total perturbances, which include the nonlinear and linear dynamics of the system, are given by $\xi_{jc} = \xi_{jc}\left(z_1, \dot{z}_1, z_{u1}, \dot{z}_{u1}, \varepsilon_1, \cdots, z_4, \dot{z}_4, z_{u4}, \dot{z}_{u4}, \varepsilon_4\right)$. The unmodeled dynamics, such as the internal and external perturbations and the parametric variations $\varepsilon_i, i = 1,2,3,4$, can also be included in this term. The coefficients d_{ij} come from the relationship between equivalent forces and forces $f_i, i = 1,2,3,4$:

$$
u_c = \begin{bmatrix} u_z \\ u_\theta \\ u_\varphi \end{bmatrix} = \begin{bmatrix} 1 & 1 & 1 & 1 \\ -a_1 & -a_1 & a_2 & a_2 \\ b_1 & -b_2 & b_1 & -b_2 \end{bmatrix} \begin{bmatrix} f_1 \\ f_2 \\ f_3 \\ f_4 \end{bmatrix}
$$

Then, we can arrive at three ultra-local equations to create three coupled MFC to control the vertical vibration z_c, the oscillatory angle θ_c and the rotational angle φ_c:

$$
\begin{cases} \ddot{z}_c = \xi_{zc} + \alpha_1 u_z \\ \ddot{\theta}_c = \xi_{\theta c} + \alpha_2 u_\theta \\ \ddot{\varphi}_c = \xi_{\varphi c} + \alpha_3 u_\varphi \end{cases}
\qquad [4.12]
$$

For the estimate of $\xi_{zc}, \xi_{\theta c}$ and $\xi_{\varphi c}$, the steps cited in the previous section are applied, giving:

$$
\begin{cases} \hat{h}_{zc} = -\dfrac{5!}{2L^5} \int_0^L \left[\left(-L^2 + 6L\tau - 6\tau^2\right) z_c(\tau) + \dfrac{\alpha_1}{2}\tau^2 (L-\tau)^2 u_z(\tau) \right] d\tau \\[3mm] \hat{h}_{\theta c} = -\dfrac{5!}{2L^5} \int_0^L \left[\left(-L^2 + 6L\tau - 6\tau^2\right) \theta_c(\tau) + \dfrac{\alpha_2}{2}\tau^2 (L-\tau)^2 u_\theta(\tau) \right] d\tau \\[3mm] \hat{h}_{\varphi c} = -\dfrac{5!}{2L^5} \int_0^L \left[\left(-L^2 + 6L\tau - 6\tau^2\right) z_\varphi(\tau) + \dfrac{\alpha_3}{2}\tau^2 (L-\tau)^2 u_\varphi(\tau) \right] d\tau \end{cases}
\quad [4.13]
$$

where L is the sliding window.

4.4.4. MFC equation

The expressions of controller equations from the system of equation [4.12] are i-PID controllers:

$$\begin{cases} u_z(t) = -\dfrac{\hat{h}_{zc}}{\alpha_1} + \dfrac{\ddot{z}_{c\,ref} - \left(K_{Pz}e_z + K_{Iz}\int e_z + K_{Dz}\dot{e}_z\right)}{\alpha_1} \\[4mm] u_\theta(t) = -\dfrac{\hat{h}_{\theta c}}{\alpha_2} + \dfrac{\ddot{\theta}_{c\,ref} - \left(K_{P\theta}e_\theta + K_{I\theta}\int e_\theta + K_{D\theta}\dot{e}_\theta\right)}{\alpha_2} \\[4mm] u_\varphi(t) = -\dfrac{\hat{h}_{\varphi c}}{\alpha_3} + \dfrac{\ddot{\varphi}_{c\,ref} - \left(K_{P\varphi}e_\varphi + K_{I\varphi}\int e_\varphi + K_{D\varphi}\dot{e}_\varphi\right)}{\alpha_3} \end{cases} \qquad [4.14]$$

The controllers developed in this chapter have the purpose of attenuating vertical vibrations and oscillatory and rotational fluctuations at the same time. By basing it on three ultra-local models, we deduced three laws for intelligent control. Three reference trajectories are required in order to produce the weakest fluctuations and efficiently regulate the rotational and oscillatory angles of the vehicle's body.

In this study, the configuration of sensors was chosen such that we could estimate the perturbances at a low cost by using the available sensors installed in the vehicle. In the front tires, two accelerometers were installed to measure accelerations $\ddot{z}_{u1}$ and $\ddot{z}_{u2}$. A gyroscopic sensor was installed to collect the oscillatory and rotational speeds. Finally, an accelerometer was installed to measure the acceleration of the body of the vehicle.

These algebraic estimators include a process for reducing noise and were experimentally verified to be efficient in resolving a variety of control problems covering a wide array of applications. When calculating the estimate, the presence of the integral acts like a low-pass filter, which naturally reduces the effects of noise, making it possible to arrive at a good estimate of the derivative of the measured signal. In practical applications, the length of the sliding window L should be in the interval [20.Ts–100.Ts] (Maalej 2014).

4.4.5. *Numerical simulation*

Notation	Value
$a_1 \ (m)$	1.011
$a_2 \ (m)$	1.803
$b_1 \ (m)$	0.761
$b_2 \ (m)$	0.755
$m_c \ (kg)$	1460
$m_{u1,2} \ (kg)$	40
$m_{u3,4} \ (kg)$	35.5
$k_{u(1,2,3,4)} \ (N/m)$	175,500
$k_{(1,2)} \ (N/m)$	19,960
$k_{(3,4)} \ (N/m)$	17,500
$d_{(1,2)} \ (Ns/m)$	1,290
$d_{(3,4)} \ (Ns/m)$	1,620
$I_x \ (kg.m^2)$	460
$I_y \ (kg.m^2)$	2,460

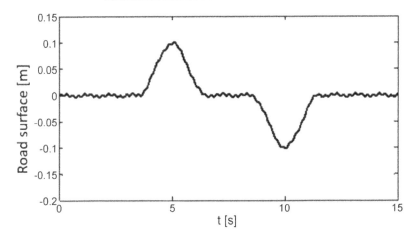

Figure 4.3. *The road surface used*

In the presence of nonlinearities, the qualitative performance of suspension systems worsens relative to the comfort of passengers. In fact, a proposition to demonstrate the efficiency of the MFC controller in rejecting this kind of problem was studied. The type of stimulus considered in this study consisted of specific irregularities (see Figure 4.2). The specific irregularities are described as bumps or potholes (Haddar et al. 2019b):

$$z_r(t) = \begin{cases} -0.0592\, t_a^3 + 0.1332\, t_a^2 + f(t); 3.5 \leq t \leq 5 \\ 0.0592\, t_b^3 + 0.1332\, t_b^2 + f(t); 5 \leq t \leq 6.5 \\ 0.0592\, t_c^3 - 0.1332\, t_c^2 + f(t); 8.5 \leq t \leq 10 \\ -0.0592\, t_d^3 - 0.1332\, t_d^2 + f(t); 10 \leq t \leq 11.5 \end{cases}$$ [4.15]

where $f(t)$ represents a sinusoidal function equal to $0.002\sin(2\pi t) + 0.002\sin(7.5\pi t)$; and t_a, t_b, t_c and t_d correspond to the following time intervals: $t_a = t - 3.5, t_b = t - 6.5, t_c = t - 8.5,\ t_d = t - 11.5$.

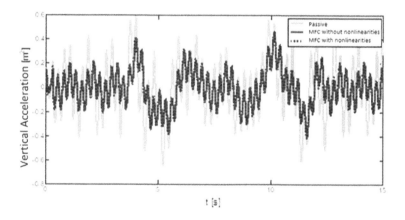

Figure 4.4. *Vertical acceleration of the center of gravity. For a color version of this figure, see www.iste.co.uk/elhami/uncertainty.zip*

The results of the numerical simulations using MATLAB/SIMULINK for a vehicle driving over the surface are shown in Figure 4.3. Figures 4.4 and 4.5 show the vertical acceleration of the vehicle's center of mass and the acceleration of the suspended mass at each corner of the vehicle. As is well known, the acceleration of the vehicle directly determines the comfort of the passengers. We can clearly see that with respect to passive suspension, the insulation effect from the vibrations of the active suspension is very significant and the performance of the proposed controller is better.

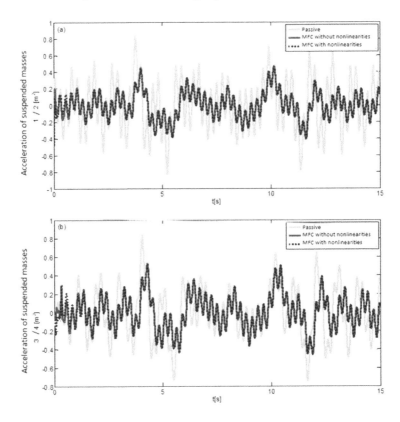

Figure 4.5. *Accelerations of suspended masses without active control and with MFC. For a color version of this figure, see www.iste.co.uk/elhami/uncertainty.zip*

In order to ensure the safety of driving and the stability of the grip on the road, the dynamic load on the tire must be inferior to its static load. The normalized amplitude of deformation of the tire for each side must be inferior to a unit. According to the results illustrated in Figure 4.6, this value is inferior to 1 and further improved when using MFC.

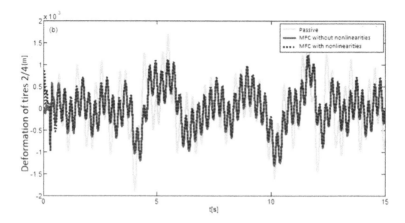

Figure 4.6. *Normalized deformations of tires. For a color version of this figure, see www.iste.co.uk/elhami/uncertainty.zip*

Figures 4.7 and 4.8 show the accelerations of a vehicle with active and passive suspension in two directions: oscillatory and rotational. MFC has a good ability to attenuate the fluctuations of rotation and oscillation.

The four control forces in MFC are shown in Figure 4.9. We can see that none of the control forces surpass 10^3 N, which means they are rather small and can be easily obtained by the force actuators generally used in active suspension.

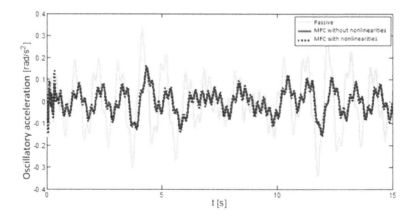

Figure 4.7. *Oscillatory acceleration without an active controller and with MFC. For a color version of this figure, see www.iste.co.uk/elhami/uncertainty.zip*

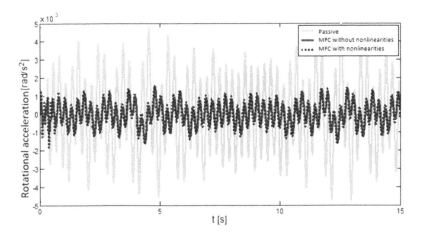

Figure 4.8. *Rotational acceleration without an active controller and with MFC. For a color version of this figure, see www.iste.co.uk/elhami/uncertainty.zip*

The integrated algebraic estimator in each control equation can precisely estimate the unknown perturbance that comes from the nonlinearities. The simulations show the ability of an intelligent MFC controller to work properly with a nonlinear system and to manage changes without requiring the recalibration of parameters in the controller.

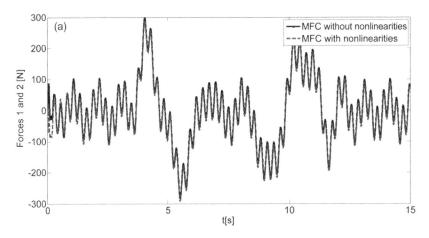

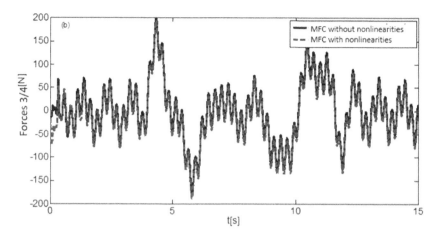

Figure 4.9. *Forces generated by the controllers. For a color version of this figure, see www.iste.co.uk/elhami/uncertainty.zip*

4.5. Conclusion

In this chapter, an intelligent method was developed based on MFC for the comfort of driving a vehicle with a complete active suspension system. In the proposed controller, algebraic estimators were created to estimate the total perturbation in three directions (vertical, oscillatory and rotational), including external stimulus from the road, the nonlinearities of the system and modeling errors, if they exist. A direct method does not require an overall mathematical model with easy implementation. The intelligence of the controller can be summarized by its capacity to make decisions quickly and online, in fractions of a second, and to reject all sorts of perturbations without needing to know their models or frequential characteristics.

4.6. References

Andary, S., Chemori, A., Benoit, M., Sallantin, J. (2012). A dual model free control of under actuated mechanical systems – Application to the inertia wheel inverted pendulum with real-time experiments. In *2012 American Control Conference (ACC)*, IEEE, Montreal, 1029–1034.

Bouazara, M., Richard, M.J., Rakheja, S. (2006). Safety and comfort analysis of a 3-D vehicle model with optimal non-linear active seat suspension. *Journal of Terramechanics*, 43(2), 97–118.

Fergani, S., Menhour, L., Sename, O., Dugard, L., Novel, B.A. (2014). Full vehicle dynamics control based on LPV/$\mathcal{H}\infty$ and flatness approaches. In *2014 European Control Conference (ECC)*, IEEE, 2346–2351.

Fliess, M. and Join, C. (2008). Intelligent PID controllers. In *2008 16th Mediterranean Conference on Control and Automation*, IEEE, 326–331.

Fliess, M. and Join, C. (2013). Model-free control. *International Journal of Control*, 86(12), 2228–2252.

Fliess, M. and Join, C. (2020). Machine learning and control engineering: The model-free case. In *Proceedings of the Future Technologies Conference*, 258–278. Springer, Cham.

Fliess, M. and Sira-Ramírez, H. (2003). An algebraic framework for linear identification. *ESAIM: Control, Optimisation and Calculus of Variations*, 9, 151–168.

Fliess, M. and Sira-Ramirez, H. (2004). Control via state estimations of some nonlinear systems. *6th IFAC Symposium on Nonlinear Control Systems 2004 (NOLCOS 2004)*, Stuttgart, 1–3 September, 37(13), 41–48.

Fliess, M., Lévine, J., Martin, P., Rouchon, P. (1995). Flatness and defect of non-linear systems: Introductory theory and examples. *International Journal of Control*, 61(6), 1327–1361.

Fliess, M., Join, C., Riachy, S. (2011). Nothing is as practical as a good theory: Model-free control. Rien de plus utile qu'une bonne théorie : la commande sans modèle. *Mathematics: Optimization and Control*, arXiv:1103.5897. doi: 10.48550/arXiv.1103.5897.

Haddar, M., Baslamisli, S.C., Chaari, R., Chaari, F., Haddar, M. (2019a). Road profile identification with an algebraic estimator. *Proceedings of the Institution of Mechanical Engineers, Part C: Journal of Mechanical Engineering Science*, 233(4), 1139–1155.

Haddar, M., Chaari, R., Baslamisli, S.C., Chaari, F., Haddar, M. (2019b). Intelligent PD controller design for active suspension system based on robust model-free control strategy. *Proceedings of the Institution of Mechanical Engineers, Part C: Journal of Mechanical Engineering Science*, 233(14), 4863–4880.

Haddar, M., Chaari, R., Baslamisli, S.C., Chaari, F., Haddar, M. (2021). Intelligent optimal controller design applied to quarter car model based on non-asymptotic observer for improved vehicle dynamics. *Proceedings of the Institution of Mechanical Engineers, Part I: Journal of Systems and Control Engineering*, 235(6), 929–942.

Karnopp, D. (1995). Active and semi-active vibration isolation. In *Current Advances in Mechanical Design and Production VI*, Elarabi, M.E. and Wifi, A.S. (eds). Pergamon Press, Oxford.

Maalej, S (2014). Commande robuste des systèmes à paramètres variables. PhD Thesis, Lille 1 University of Science and Technology, Lille.

Menhour, L., d'Andréa-Novel, B., Fliess, M., Gruyer, D., Mounier, H. (2015). A new model-free design for vehicle control and its validation through an advanced simulation platform. In *2015 European Control Conference (ECC)*, IEEE, 2114–2119.

Ni, T., Li, W., Zhao, D., Kong, Z. (2020). Road profile estimation using a 3D sensor and intelligent vehicle. *Sensors*, 20(13), 3676.

Othmane, A., Kiltz, L., Rudolph, J. (2022). Survey on algebraic numerical differentiation: Historical developments, parametrization, examples, and applications. *International Journal of Systems Science*, 53(9), 1848–1887.

Polack, P. (2018). Consistency and stability of hierarchical planning and control systems for autonomous driving. PhD Thesis, PSL Research University, Paris.

Ren, W. and Beard, R.W. (2008). *Distributed Consensus in Multi-Vehicle Cooperative Control*, 1st edition. Springer, London.

Shafie, A.A., Bello, M.M., Khan, R.M. (2015). Active vehicle suspension control using electro hydraulic actuator on rough road terrain. *Journal of Advanced Research*, 9(1), 15–30.

Sidhom, L. (2011). Sur les différentiateurs en temps réel : algorithmes et applications. PhD Thesis, INSA de Lyon, Lyon.

Valášek, M., Novak, M., Šika, Z., Vaculin, O. (1997). Extended ground-hook – New concept of semi-active control of truck's suspension. *Vehicle System Dynamics*, 27, 289–303.

Villagra, J. and Balaguer, C. (2009). An algebraic approach for accurate motion control of humanoid robot joints. In *International Conference on Intelligent Robotics and Applications*, 723–732. Springer, Berlin, Heidelberg.

Yosida, K. (2012). *Operational Calculus: A Theory of Hyperfunctions*. Springer, New York.

Zhang, X., Yu, W., Ma, F., Zhao, F., Guo, K. (2013). Semi-active suspension adaptive control strategy based on hybrid control. In *Proceedings of the FISITA 2012 World Automotive Congress*. Springer, Berlin, Heidelberg.

5

Optimization of the Power Inductor of a DC/DC Converter

In this chapter, we are interested in the study of power inductors, which are very important elements in the DC/DC converter. The performance of the converters is therefore a function of that of the inductors. In the same way, during a commissioning test, the ferrite of the inductor breaks, which represents a challenge for developers and manufacturers. After the thermomechanical modeling of the system by the finite element method using the Comsol Multiphysics software, it was found that the equivalent stresses within the ferrite exceed the value of the limit of elasticity which leads to the rupture of the ferrite. Therefore, we will develop an optimization approach by the KA-CMA-ES method, to minimize the effect of equivalent stresses in the structure. This approach is based on the coupling between two models: the finite element model with Comsol Multiphysics and the optimization model using the software Matlab. After the application of our approach, we succeeded in minimizing the stresses equivalent to a value less than the elastic limit and consequently the correct operation of the inductors during the entire test period.

5.1. Introduction

The DC/DC converter is an important element in the mechanisms of electric or hybrid vehicles. It is a system of connection and adaptation between the power sources and the load; it also allows us to improve the voltage gain, the efficiency and the capacity of the power handling (Wang et al. 2019). The DC/DC converter is mainly composed of a transformer and two inductors; it is placed under the engine cover of the vehicle to convert the power received from the battery to another power compatible with the electrical equipment of the vehicle.

The two inductors are composed of a coil and a ferrite; the coil is bonded to the ferrite by a thermally insulated glue. The performance and reliability of these converters is an issue for developers and manufacturers today. While the operating and environmental conditions influence the reliability of the component (Bendaou et al. 2017). This results in failures of different types: electrical, thermal and mechanical. Indeed, during a test of the commissioning of the converter under the operating conditions: electrical loads and thermal loads, a mechanical problem appears in the ferrite of the inductor after the duration of the test, which represents a great impact on the proper functioning of the system. Therefore, it is necessary to optimize its reliability and performance.

One of the optimization methods is the CMA-ES (Covariance Matrix Adaptation-Evolution Strategy) method, which is a stochastic research method based on the population in continuous and discrete spaces (Huang et al. 2018) and is used to solve optimization problems by minimizing an objective function. The objective function to be minimized is a function of failure mode. This method has shown efficiency through the results obtained in several works aimed at the reliability of systems (El Hami and Pougnet 2015). The CMA-ES method is based on the finite element model to solve the optimization problem, but this method is no longer practical when the finite element model lasts tens of minutes. However, the optimization process requires hundreds or thousands of simulations. By integrating the CMA-ES method with the kriging metamodel, we obtain the KA-CMA-ES method (Kriging Assisted-Covariance Matrix Adaptation-Evolution Strategy) which makes it possible to overcome the time constraint of the calculation (Huang et al. 2017).

In order to solve this industrial problem, we will treat the reliability of the converter with a different approach. In order to identify the causes of this failure, we develop a finite element model with the Comsol Multiphysics software; this model makes it possible to simulate the Multiphysics behavior of the inductor. Thereafter, we develop an optimization approach based on the coupling of two models: the finite element model with the Comsol software and the optimization model based on the KA-CMA-ES method. This approach will allow us to minimize the effect of mechanical failure on the structure of the inductor.

5.2. Description of the power inductor

The DC/DC converter has been used in several industrial applications. In our case, we are interested in the converter used in electric vehicles which ensures the adaptation of the power supply of their electronic equipment. It is composed of a transformer and two power conductors (Bendaou 2017). Inductors are composed of a ferrite and a coil; they produce a magnetic field and induce a voltage in themselves if they are traversed by a current. Moreover, they make it possible to store and release energy during the different operating states of the circuit (Ragusa et al. 2020). The element that caught our attention is ferrite; its use is to take advantage of its magnetic and electrical properties. However, this element has mechanical drawbacks such as its fragility and sensitivity to the temperature gradient. In the following, we will study the influence of its drawbacks on the reliability of the component.

Figure 5.1. *Power inductor structure. For a color version of this figure, see www.iste.co.uk/elhami/uncertainty.zip*

5.3. Thermomechanical modeling of the power inductor

Finite element modeling makes it possible to simulate the multi-physical behavior of systems. In our case, the thermomechanical modeling of the inductor allows us to observe its thermal and mechanical behavior according to the loadings and the boundary conditions applied. The resolution of the finite element model will be done by the Comsol Multiphysics software. The properties of the materials necessary for the realization of the model are classified in Table 5.1.

Materials	Aluminum	Glue	Ferrite	Copper
K	235	1.7	5	400
α	1.66e-5	4.45 e-5	7.8 e-6	1.66 e-5
E	7.3 e10	3.54 e9	150 e6	1.14e11
Poisson's ratio	0.33	0.3	0.28	0.355

Table 5.1. *Thermal properties of materials*

5.3.1. *Thermal modeling*

The current intensity, which crosses the coil of the inductor, is 50 A. This current induces thermal losses at the level of the coil characterized by a resistance of 0.82 momega. A heat loss was also observed at the level of the 0.7 W ferrite. To study the thermal behavior of the power conductor, it is necessary to solve the following heat equation:

$$k.\nabla^2 T = \rho \cdot C_{p.} \frac{\partial T}{\partial t} + Q \qquad [5.1]$$

with:

– Q: the power dissipated in joules;

– K: the thermal conductivity $(W.m^{-1}.K^{-1})$;

– ρ: the density $(Kg.m^{-3})$;

– C_p: the heat mass capacity $(J.Kg^{-1}.K^{-1})$;

– T: the temperature (K).

The resolution of this equation is performed using the Comsol Multiphysics software. After the application of the loads and the boundary conditions: thermal losses in the coil (2.5 W) and the ferrite (0.7 W), the ambient temperature 25°C and the cooling temperature 22°C.

Figure 5.2 shows the distribution of the temperature in the structure of the inductor. The temperature is very important above the ferrite; on the other hand, it is minimal below the ferrite thanks to the cooling system.

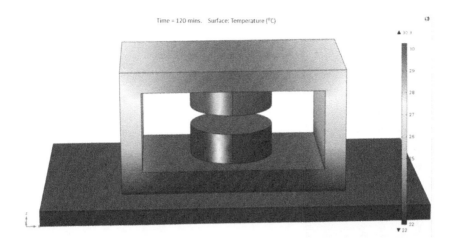

Figure 5.2. *Distribution of the temperature in the structure of the power inductor. For a color version of this figure, see www.iste.co.uk/elhami/uncertainty.zip*

In addition, Table 5.2 presents the comparison of simulation and experimental results. According to this table, the relative error between the results is due to several factors such as the uncertainties of the properties of the materials, the geometric reproduction and the sensitivity of the means of measurement. We also note that the value of the relative error is less important and acceptable. This allows us to validate the finite element model developed for the thermal modeling of the inductor.

Type of test		Numerical	Experimental
Bobine	Tmin °C	34.5	38
	Tmax °C	53.5	57
Ferrite	Tmin °C	24	25
	Tmax °C	31	27

Table 5.2. *Numerical simulation and experimental results*

In order to observe the evolution of the temperature as a function of the Ferrite structure, Figure 5.3 shows that the temperature increases as a function of the test time following a nonlinear evolution. Its value stabilizes from 90 min of the test duration.

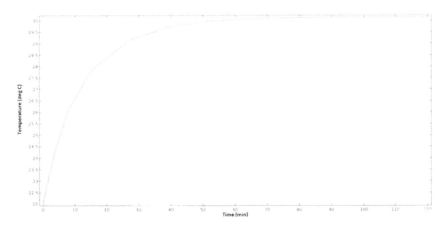

Figure 5.3. *Evolution of the temperature as a function of time. For a color version of this figure, see www.iste.co.uk/elhami/uncertainty.zip*

5.3.2. *Thermomechanical modeling*

The application of electrical and thermal charges on the inductor also induces mechanical phenomena in its structure. In order to identify these phenomena, we will develop a thermomechanical model thanks to the coupling of the thermal model and the mechanical model, which can be translated in the form of equations by the equations of thermo-elasticity. We can define the deformations and the density of entropy by the following expressions (El Hami and Pougnet 2015):

$$\{\varepsilon\} = [D]^{-1}\{\sigma\} + \{\alpha\}\Delta T \tag{5.2}$$

$$\{S\} = \{\alpha\}^T\{\sigma\} + \frac{\rho \cdot C_P}{T_0}\Delta T; \Delta T = T - T_{ref} \tag{5.3}$$

with $\{\varepsilon\}$ being the vector of total deformation. It is a function of the vector of the constraints, the vector of the coefficients of thermal dilation, the difference of the temperature and the inverse of the matrix of elastic rigidity.

In equation [5.3], the entropy is a function of several thermal and mechanical parameters such as stresses and temperature variation and others.

In this section, we will present the results of thermomechanical modeling of the inductor under the same loads for the thermal part. Figure 5.4 shows the distribution of displacements in the structure of the ferrite. The displacements are greater in the upper part of the ferrite and can reach approximately 3 µm.

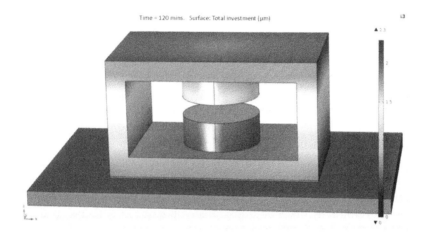

Figure 5.4. *Distribution of the displacement in the structure. For a color version of this figure, see www.iste.co.uk/elhami/uncertainty.zip*

Table 5.3 presents a comparison of the simulation results with that of the experimental. We can note that the relative error is less important also for this modeling.

Type of test		Numerical	Experimental
Bobine	Dmin (µm)	0	0
	Dmax (µm)	2.98	2.7035

Table 5.3. *The physical properties of materials*

According to these two modelings, we note that the model finite element is able to simulate the thermomechanical behavior of the inductor. This model will be used to solve the optimization problem that we developed in the rest of this chapter.

5.4. Optimization methods

In this section, we will detail the optimization methods that we used. We will start with the definition of the CMA-ES method and its parameters. Thereafter, we will talk about the same method assisted by the kriging metamodel, its management and quality control and finally the global KA-CMA-ES method.

5.4.1. *Covariance matrix adaptation-evolution strategy (CMA-ES)*

The CMA-ES (Covariance Matrix Adaptation-Evolution Strategy) method is the best member of the family of evolution algorithms. It is a derivative-free stochastic optimization method and can adapt to non-convex optimization problems. CMA-ES is based on the four principles of natural selection: evolution, selection, recombination and mutation. The recombination of the elements (parents) of a population generates new individuals (children); the new individuals undergo a mutation so that they become new members of the same population, in order to keep and adapt the population size; it is necessarily making the selection by choosing the best individuals after the mutation. The implementation of evolutionary operations ensures this approach, to guide the approach towards a global optimum in a continuous or discrete search space (Hamdani et al. 2019). CMA-ES(λ,μ) is based on the adaptation of the covariance matrix of the multinormal law in Rn. This adaptation is equivalent to the construction of an approximation of the objective function f. This method generates one new individual from μ elements of the population. CMA-ES(λ,μ) generates the population by sampling a multinormal distribution (Amar et al. 2022):

$$x_k^{(g+1)} \sim m^{(g)} + \sigma^{(g)} \mathcal{N}\left(0, C^{(g)}\right) \quad \text{for} \quad k = 1,\ldots,\lambda \qquad [5.4]$$

where:

– $\sim$ indicates the same distribution on the left and on the right;

– $\mathcal{N}\left(0, C^{(g)}\right)$ is a multivariate normal distribution with zero mean and covariance matrix;

– $x_k^{(g+1)} \in \mathbb{R}^n$ is the kth offspring (individual, research point) of generation g+1;

– $m^{(g)} \in \mathbb{R}^n$ is the average value of the distribution of research at generation g;

– $\sigma^{(g)} \in \mathbb{R}_{>0}$ is the step size, to generation g. C(g) $\in \mathbb{R}n\times n$, covariance matrix at generation g;

– $\lambda \geq 2$ is the population size.

As a result, the images of the individuals generated by sampling are evaluated by the objective function. The best individuals will be grouped together in the average vector m, which has the following expression:

$$m^{(g+1)} = \sum_{i=1}^{\mu} w_i x_{i:\lambda}^{(g+1)} \qquad [5.5]$$

and

$$\sum_{i=1}^{\mu} w_i = 1, \quad w_1 \geq w_2 \geq \cdots \geq w_\mu > 0 \qquad [5.6]$$

where:

– $\mu \leq \lambda$ is the number of selected points or the size of the original population (parents);

– $w_{i=1\ldots\mu} \in \mathbb{R}_{>0}$ represents the recombination peas, which are strictly positive and verify equation [5.6];

– $x_{i:\lambda}$ is the best ith individual selected from the set of individuals $[x_1, x_2, \ldots, x_\lambda]$ generated by equation [5.4];

– "i, λ" is the ith index of the individual after ranking such that $f(x_{1:\lambda}) \leq f(x_2 \cdot \lambda) \leq \cdots \leq f(x_{\lambda:\lambda})$, where f is the objective function.

After the update of the average vector m, the next step is the step adaptation σ and the covariance matrix C. The update of the two parameters is done by two evolution paths and Huang et al. (2018). However, the rest of the process of the CMA-ES method requires updating the evolution path, allowing us to update the size of the step σ. The two parameters are calculated according to the following equations:

$$p_{\sigma}^{(g+1)} = (1-c_{\sigma}) p_{\sigma}^{(g)} + \sqrt{c_{\sigma}(2-c_{\sigma})} \sqrt{\mu_w} \left(C^{(t)}\right)^{-\frac{1}{2}} \frac{m^{(g+1)} - m^{(g)}}{\sigma^{(t)}}$$

$$= (1-c_{\sigma}) p_{\sigma}^{(g)} + \sqrt{c_{\sigma}(2-c_{\sigma})} \sqrt{\mu_w} \sum_{i=1}^{\mu} w_i z_{i\lambda}$$

[5.7]

$$\sigma^{(g+1)} = \sigma^{(g)} \exp\left(\frac{c_{\sigma}}{d_{\sigma}} \left(\frac{\left\|p_{\sigma}^{(g+1)}\right\|}{\mathbb{E}\left(\left\|\mathcal{N}_d(0,I)\right\|\right)} - 1\right)\right)$$

[5.8]

where:

– $p_{\sigma}^{(g)} \in \mathbb{R}^d$ is the evolution path for the step size at generation g; its initial value is $p_{\sigma}^{(0)} = 0$;

– $c_{\sigma} \in [0,1]$ is the time constant for step size adaptation, whose default value is $c_{\sigma} = (\mu_w + 2) / (d + \mu_w + 5)$;

– μ_w is called the variance effective selection mass, whose value is

$$\mu_w = \left(\sum_{i=1}^{\mu} w_i\right)^2 / \sum_{i=1}^{\mu} w_i^2.$$

To update the covariance matrix C, it is necessary to update the path of evolution p_c, which is performed according to the following equations:

$$p_c^{(g+1)} = (1-c_c) p_c^{(g)} + h_{\sigma} \sqrt{c_c(2-c_c)} \sqrt{\mu_u} \frac{m^{(g+1)} - m^{(g)}}{\sigma^{(g)}}$$

$$= (1-c_c) p_c^{(g)} + h_{\sigma} \sqrt{c_c(2-c_c)} \sqrt{\mu_w} \sum_{i=1}^{\mu} w_i \left(C^{(g)}\right)^{\frac{1}{2}} z_{i:\lambda}$$

[5.9]

While the covariance matrix becomes as follows:

$$C^{(g+1)} = \left(1 - c_1 - c_u\right)C^{(g)} + c_1\left(p_c^{(g+1)}\left(p_c^{(g+1)}\right)^T + \delta(h_\sigma)C^{(g)}\right)$$

$$+ c_\mu \sum_{i=1}^{u} w_i \left(\left(C^{(g)}\right)^{\frac{1}{2}} z_{i,\lambda}\right)\left(\left(C^{(g)}\right)^{\frac{1}{2}} z_{i:\lambda}\right)^T \qquad [5.10]$$

where:

– $p_c^{(g)} \in \mathbb{R}^d$ is the evolution path for step size at generation g; its initial value is $p_c^{(0)} = 0$;

– $c_c \in [0,1]$ is the adaptation constant of the covariance matrix;

– h_σ is the Heaviside function;

– c_1 and c_μ express the learning rate, respectively, for rank-one-update and μ-update $\left(C^{(t)}\right)^{\frac{1}{2}}$, their default values are $c_1 = \dfrac{2}{(d+1.3)^2 + \mu_{uv}}$ and

$$c_u = \min\left(1 - c_1, 2\frac{\mu_v - 2 + 1/\mu_w}{(d+2)^2 + \alpha_u\mu_w / 2}\right) \text{ with } \alpha_\mu = 2;$$

and

$$\delta(h_\sigma) = (1 - h_\sigma)c_c(2 - c_c)$$

All of these steps can be organized in an algorithm, which is executed according to the chosen convergence conditions following a very specific order. Algorithm 5.1 is repeated until the convergence condition is satisfied by returning the best individual (Hansen 2016).

5.4.2. *Covariance matrix adaptation-evolution strategy assisted by kriging (KA-CMA-ES)*

To carry out an optimization study using the CMA-ES method, it is necessary to carry out several (thousands of) simulations using the finite element model, and each simulation can last more than tens of minutes, which is costly in terms of time. However, consideration was given to integrating the kriging metamodel into the CMA-ES algorithm using the approximate ranking procedure as a quality control method for the metamodel. The CMA-ES algorithm will retain all its main instructions, except that the image evaluation step can be done by the metamodel. In the second case, the approximate ranking procedure is used, if n_k the number of points required to construct the metamodel is greater than the number of points T in the training pose. This integration of the metamodel into the evolutionary method can be expressed in the form of an algorithm (Algorithm 5.1) called KA-CMA-ES (Kriging-assisted CMA-ES) (Hamdani et al. 2019).

Algorithm 5.1. Kriging-assisted CMA-ES

Initialize evolution path $p_s^{(0)} = 0, p_c^{(0)} = 0,$ the covariance matrix $C^{(i)} = I$ the step size a^{ion} and the selection parameters.

Initialize the mean vector $m^{(0)}$ to a random candidate $g - 0$

while the convergence criterion is not reached do:

$T \leftarrow \left\{ (x_i, f_i) \in S \;\|\; (x_i - m^{(b)})^T \right.$

$\left. (((\infty)^2 c^4) - 1 (x_i - m^{(y)}) \le r^2 \right\}$

if $|T| < m_2$ then

evaluation

$f_k - f(x_k^{(s+1)}) k = 1, \ldots, \lambda$

$S = S \cup \left\{ (x_k^{(9+1)}, f_k) \right\}_{\varepsilon - 1}^{\lambda}$

$\leftarrow \left\{ (x_i, f_i) \in | (x_i - m^{(g)})^T ((\sigma^{(g)})^2 C^{(\Delta)})^{-1} (x_i - m^{(g1)}) \le r^2 \right\}$

else \\

$\mathbb{R}$ Metamodel-based evaluation

Run approximate ranking according to Algorithm 5.1

end if

$m^{(g+1)} = \sum_{i=1}^{\mu} w_i x_{i=\lambda}^{(s)}$

$p_\sigma^{(g+1)} = (1 - c_\sigma) p_\sigma^{(g)} + \sqrt{c_\sigma (2 - c_\sigma) \mu_{eff}} (C^{(g)})^{-\frac{1}{2}} \dfrac{m^{(g+1)} - m^{(g)}}{\sigma^{-(g)}}$

$$\sigma^{(g+1)} = \sigma^{(g)} \exp\left(\frac{c_\sigma}{d_\sigma}\left(\frac{\left\|p_\sigma^{(g+1)}\right\|}{E\left(\left\|N(0,I)\right\|\right)} - 1\right)\right)$$

$$p_c^{(g+1)} = (1 - c_c)\,p_c^{(g)} + h_\sigma\,\sqrt{c_c\,(2 - c_c)\,\mu_{eff}}\;\frac{m^{(g+1)} - m^{(g)}}{\sigma^{(g)}}$$

$$C^{(g+1)} = \left(1 - c_1 - c_\mu + c_1\delta(h_\sigma)\right)C^{(g)} + c_1\,p_c^{(g+1)}\left(p_c^{(g+1)}\right)^T \ldots$$

$$+c_\mu \sum_{i=1}^{u} w_i\left(\frac{x_{i:\lambda}^{(g+1)} - m^{(g)}}{\sigma^{(g)}}\right)\left(\frac{x_{i=1}^{(g+1)} - m^{(g)}}{\sigma^{(g)}}\right)^T$$

end while

return best optimum x_{opt}

5.5. Optimization of the power inductor

5.5.1. *Description of the optimization problem*

In order to test the performance of the DC/DC converter, tests are run under the operating conditions in cars. After 120 min of testing, the ferrite of the inductor breaks, which presents a challenge in terms of the reliability of the entire component. To study and analyze this phenomenon, we used the thermomechanical model previously developed under the following operating conditions: an ambient temperature of 105°C which represents the temperature under the engine cover, and a water-cooling temperature of 75°C.

The temperature distribution at the ferrite is shown in Figure 5.5. It can be noted that the temperature is always at maximum above the ferrite.

For the purpose of identifying the causes of the conductor ferrite breakage under test conditions that represent actual operating conditions, we made numerical simulations of the von Mises equivalent stresses using the same thermomechanical model and under the same boundary conditions. Figure 5.6 shows the distribution of these stresses within the structure of the ferrite. It can be clearly noted that the maximum value is 56.4 MPa below the ferrite. Therefore, it can be seen that the cause of failure is that the maximum von Mises stresses exceed the yield strength of 50 MPa defined by the manufacturer.

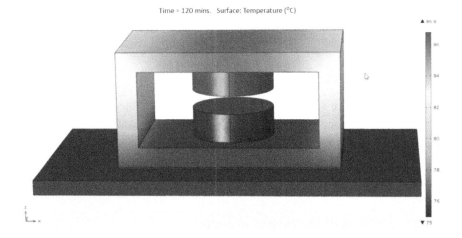

Figure 5.5. *Temperature distribution in the structure of the power inductor after the test. For a color version of this figure, see www.iste.co.uk/elhami/uncertainty.zip*

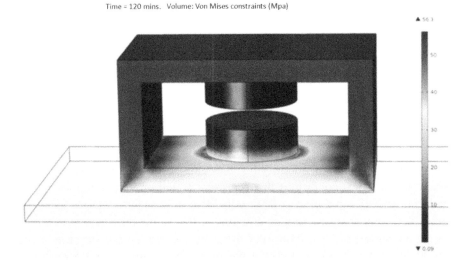

Figure 5.6. *Distribution of von Mises stresses in the structure after optimization. For a color version of this figure, see www.iste.co.uk/elhami/uncertainty.zip*

To solve this problem, we minimized the effect of von Mises stresses on the ferrite structure. The optimization problem under the given constraints can be expressed in the following form:

$$\min f(x) = \min(\sigma(x))$$
$$42 \leq x_1 \leq 50$$
$$22 \leq x_2 \leq 28$$

[5.11]

with $f(x)$ being the objective function that represents the equivalent stresses $\sigma_e(x)$ to be minimized. x_1 and x_2 are the geometric parameters that represent the width and the external length of the ferrite, respectively. On the other hand, we must keep the width and the internal length constant so as not to influence the location of the coil.

5.5.2. Results and discussion

In order to solve the optimization problem defined above, we developed an approach based on the coupling between two models: the finite element model developed with the Comsol Multiphysics software and the KA-CMA-ES optimization model coded on the Matlab software. The first model makes it possible to evaluate the value of the objective function to be minimized, and the second is used to optimize the objective function according to the constraints posed. The results obtained are listed in Table 5.4.

Parameters	Initial value	KA-CMA-ES	CMA-ES
x_1 (mm)	47	44.23	44.30
x_2 (mm)	24	25.69	25.96
σ (MPa)	56.4	40.8	40.8
t(s)		79426	188368

Table 5.4. Results of the optimization by the two methods

The KA-CMA-ES method made it possible to minimize the effect of von Mises stresses in the ferrite structure by taking into account the constraints of the geometric parameters. The maximum stresses in the new structure are 40.8 MPa; this value is lower than the elastic limit defined by the manufacturer. Figure 5.7 shows the comparison of the equivalent stress distribution in the old structure and the optimized structure.

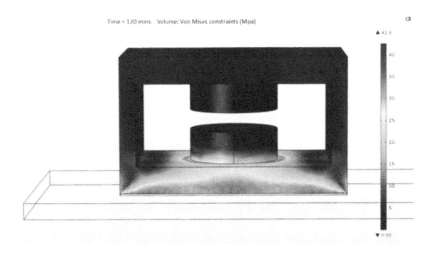

Time = 120 mins.   Volume: Von Mises constraints (Mpa)

Figure 5.7. *Distribution of von Mises stresses in the structure after optimization. For a color version of this figure, see www.iste.co.uk/elhami/uncertainty.zip*

The approach developed based on the KA-CMA-ES method has solved this industrial problem. Therefore, the breakage of the inductor ferrite can be avoided by introducing small changes in its geometry. This will ensure better performance of the component.

5.6. Conclusion

In this chapter, we have dealt with a problem from the industry in the automotive field. It is an inductor that breaks during the duration of the test under the conditions of the function in the car: ambient temperature, cooling temperature and heat losses. We modeled the component using the Comsol Multiphysics software based on the finite element method. This modeling made it possible to simulate the operation of the inductor and to identify the main cause of the breakage of its ferrite; this was because of the equivalent stresses of von Mises, which exceed the elastic limit defined by the manufacturer. To solve this problem, we have developed an approach based on the coupling between the finite element model and the optimization model based on the KA-CMA-ES method coded on the Matlab software.

This approach made it possible to determine a new optimal structure by optimizing the equivalent stresses while avoiding the fracture of the ferrite of the indicator. The proposed optimization method has shown its effectiveness in solving this industrial problem.

5.7. References

Amar, A., Radi, B., El Hami, A. (2022). Optimization based on electro-thermomechanical modeling of the high electron mobility transistor (HEMT). *International Journal for Simulation and Multidisciplinary Design Optimization*, 13. doi: 10.1051/smdo/2021035.

Bendaou, O. (2017). Caractérisation thermomécanique, modélisation et optimisation fiabiliste des packages électroniques. PhD Thesis, Normandie Université [Online]. Available at: https://tel.archives-ouvertes.fr/tel-01760307.

Bendaou, O., Gautrelet, C., El Hami, A., Agouzoul, M. (2017). Experimental and numerical analysis of thermal and thermomechanical behavior of a power inductor accompanied by a reliability study. In *The 2nd International Conference on Materials Engineering and Nanotechnology*, 12–14 May, Kuala Lumpur, 205(1).

El Hami, A. and Pougnet, P. (2015). *Embedded Mechatronic Systems 2: Analysis of Failures, Modeling, Simulation and Optimization*. ISTE Press, London, and Elsevier, Oxford.

Hamdani, H., Radi, B., El Hami, A. (2019). Optimization of solder joints in embedded mechatronic systems via Kriging-assisted CMA-ES algorithm. *International Journal for Simulation and Multidisciplinary Design Optimization*, 10, A3. doi: 10.1051/smdo/2019002.

Hansen, N. (2016). The CMA evolution strategy: A tutorial. *arXiv:1604.00772*, 1–39.

Huang, C., El Hami, A., Radi, B. (2017). Metamodel-based inverse method for parameter identification: Elastic–plastic damage model. *Engineering Optimization*, 49(4), 633–653. doi: 10.1080/0305215X.2016.1206537.

Huang, C., Radi, B., El Hami, A., Bai, H. (2018). CMA evolution strategy assisted by Kriging model and approximate ranking. *Applied Intelligence*, 48(11), 4288–4304. doi: 10.1007/s10489-018-1193-3.

Ragusa, C., Solimene, L., Musumeci, S., de la Barriere, O., Fiorillo, F., Di Capua, G., Femia, N. (2020). Computation of current waveform in ferrite power inductors for application in buck-type converters. *Journal of Magnetism and Magnetic Materials*, 502(January), 166458. doi: 10.1016/j.jmmm.2020.166458.

Wang, H., Gaillard, A., Hissel, D. (2019). A review of DC/DC converter-based electrochemical impedance spectroscopy for fuel cell electric vehicles. *Renewable Energy*, 141, 124–138. doi: 10.1016/j.renene.2019.03.130.

Study of the Influence of Noise and Speed on the Robustness of Independent Component Analysis in the Presence of Uncertainty

This chapter presents a study on the robustness of the independent component analysis (ICA) method when estimating the road surface of a quarter vehicle. To do this, the Monte Carlo (MC) stochastic technique was used, along with the inevitable parameters of uncertainty: the mass of the vehicle, the rigidity of the spring and the shock absorption. The effect of the noise generated by wind is also considered. The convergence of the intelligent ICA method was evaluated in a comparison with real surface profiles as defined by the ISO norm with the surface profiles estimated in the context of uncertainty. The results obtained demonstrate the robustness of ICA in estimating different road surfaces.

6.1. Introduction

A road surface is one of the major factors influencing the performance of a vehicle, in particular the comfort of driving and adherence to the road. The estimation of this type of disturbance is important for understanding the vibratory behavior of the vehicle, on the one hand, and designing active control systems, on the other hand (Nodeh et al. 2021; Ben Jdidia et al.

Chapter written by Dorra BEN HASSEN, Anoire BEN JDIDIA, Mohamed Slim ABBES, Fakher CHAARI and Mohamed HADDAR.

2022). This subject has been addressed by numerous researchers. Some of them use direct methods to measure the irregularities of the surface with expensive instruments such as profilometers (Healey et al. 1977) or video cameras (Xue et al. 2020). Apart from their high costs, certain instruments cannot be used in severe environmental conditions, for example, snow and wind (Nodeh et al. 2021). Other studies have suggested estimating the surface profile of a road based on the dynamic responses of the vehicle. Let us cite the example of Fauriat et al. (2016), who proposed an algorithm based on the Monte Carlo method in order to obtain an optimal estimate of the road surface. This technique, however, requires a very long calculation time.

Recently, Haddar et al. (2019) proposed the use of an algebraic estimator of road surfaces that works in real time, but requires a very specific calibration to obtain the best estimated results. Despite the importance of these studies, only a minority of them take variations in suspension parameters into account, since they are uncertain in real scenarios. In this context, Chaabane et al. (2019) proposed a comparison between the Kalman filter estimation technique and the Independent Component Analysis method, keeping track of the variation in the suspended mass of the vehicle. The authors demonstrated that the ICA method is more robust with uncertainty than the Kalman filter and presents greater sensitivity to both variations in the suspended mass and speed of the vehicle. Thus, to have a precise estimate of the irregularities in the road surface, the uncertain parameters of the model being studied must be taken into account.

To this end, this chapter proposes the use of ICA to estimate road surfaces. Ben Jdidia et al. (2022) already used this method to estimate the road surface using the dynamic responses of the complete model of the vehicle. The efficiency of ICA in estimating the road surface has been proven by several studies (Ben Jdidia et al. 2022; Chaabane et al. 2019). The novelty of this chapter resides in the study of the robustness of this method when faced with the uncertainty of parameters (suspended mass, rigidity and shock absorption) and the effect of the wind. This study was done using the Monte Carlo algorithm. Indeed, several researchers have had recourse to this method for studying robustness. To give but one example, we can cite the study by Dammak et al. (2019), who optimized the design of a trellis using the Monte Carlo algorithm. Dammak et al. (2019) used the MC optimization technique to study the robustness of the design of a new product in a case study. In this chapter, we propose the estimation of several road surface profiles when applied to the model of a quarter car with uncertain parameters using the ICA method. The robustness of the ICA is based on the MC method. The influence of the wind on the model of the vehicle was also taken into account. The comparison of the real surface profile and the estimated surface profiles in the context of uncertainty demonstrate that ICA is a reliable method in terms of estimates and robustness when parameters vary.

6.2. The model studied

This study is a continuation of the paper published by Ben Jdidia et al. (2022). The idea is indeed to combine the artificial intelligence ICA method with the Monte Carlo technique to evaluate the estimates of road surface profiles with uncertainty. The flow chart presented in Figure 6.1 describes the steps of the research.

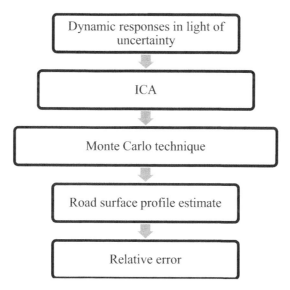

Figure 6.1. *ICA and Monte Carlo coupling*

Two road surfaces of types B and D are to be estimated with uncertainty in the different parameters of the suspension system: the variation in mass, the rigidity of the spring and the shock absorption. The Monte Carlo technique is useful in studying the robustness of ICA in estimating these road surfaces with uncertainty by calculating each iteration of relative error between the real profile and that of the ICA estimate. The number of iterations chosen was 500.

The model studied (Figure 6.1) was a quarter vehicle with two degrees of freedom, X_1 and X_2. This is the simplest model that lets us describe the vertical vibrations of the complete model (Chaabane et al. 2019). The excitation from the road surface is denoted r(t). This excitation is modeled according to ISO norm 8608. Two types of road surfaces were studied: surface B (smooth) and surface D (rough).

The equation for movement in this system can be written in the following way:

$$\begin{cases} m_1\ddot{X}_1 + k_1(X_1 - X_2) + c_1(\dot{X}_1 - \dot{X}_2) = 0 \\ m_2\ddot{X}_2 + k_1(X_2 - X_1) + c_1(\dot{X}_2 - \dot{X}_1) + k_2(X_2 - r(t)) = 0 \end{cases} \quad [6.1]$$

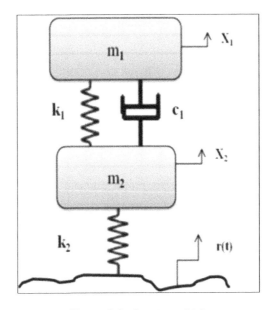

Figure 6.2. *Quarter vehicle*

The parameters used for the study of this system are presented in Table 6.1 (Fauriat et al. 2016).

Parameter	Value
m_1	372 kg
m_2	59 kg
k_1	36,540 N/m
k_2	242,000 N/m
c_1	3,300 Ns/m

Table 6.1. *Parameters for a quarter vehicle*

6.3. Construction of the road surface profile

The road surfaces studied in this chapter were constructed following the ISO norm 8608, as shown in Table 6.2.

Road surface	Degree of roughness $G_d (n_0)$ (10^{-6} m^3)		
	Lower limit	Geometric average	Upper limit
Type A	–	16	32
Type B	32	64	128
Type C	128	256	512
Type D	512	1,024	2,048
Type E	2,048	4,096	8,192

Table 6.2. *Classification of road surfaces*

To construct this road surface roughness, we used the method of integrated white noise, which considers the roughness of the road as if it were the result of a filtered white noise, presented in the following equation:

$$\dot{r}(t) = 2\pi n_0 \sqrt{G_d(n_0) \, v \, w_1(t)} \qquad [6.2]$$

where the spatial frequency N0 – 0.1 cycle/m; the spectral density of the power, $G_d(n_0)$, is given in Table 6.1; and the speed of the vehicle V = 15 m/s.

6.4. The principle of the ICA method

ICA is a form of artificial intelligence (AI) since it is in essence a black box. It is one of the methods for blind separation of sources. This problem came to light towards the end of the 1980s. It at first used in the medical field to evaluate the muscular responses stimulated by different types of stimuli. Then, the problem was developed and ended up becoming ICA (Haddar et al. 2019). This technique was largely used in several fields such as biomedical imaging, radars, medicine and, more recently, in the field of mechanics (Abbes et al. 2011; Taktak et al. 2012; Chaabane et al. 2019).

This statistical technique highlights the source signals from a mixture of measured signals. It is in essence an inverse problem that can be formulated as follows: let there be a vector $X = A.S$, where S is the source signal and A is the matrix of the mixture of measured signals. The idea consists of finding a square matrix $W = A^{-1}$ such that the vector's components are as independent as possible (Haddar et al. 2019) given the hypothesis that the estimated number of signals is equal to the number of signal sources. Figure 6.3 summarizes the concept of ICA.

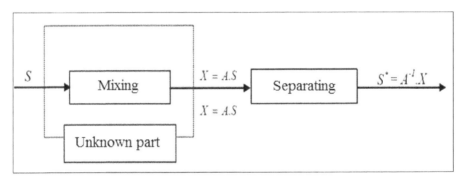

Figure 6.3. *Simplified concept of ICA (Haddar et al. 2019). For a color version of this figure, see www.iste.co.uk/elhami/uncertainty.zip*

It should be noted that the principle of separating sources is based on two principal preparatory steps, which are the centering and whitening of observed signals (Ben Jdidia et al. 2022).

– Centering

Centering consists of subtracting from vector X its average vector $m = E\{X\}$ so it has a zero average, and then S also has a zero average.

– Whitening

The second step in ICA is the whitening of observed variables. This technique makes it possible to eliminate the noise from a signal. This vector has a matrix with a covariance of one unit. We thus arrive at:

$$E\{ XX^* \} = I \tag{6.3}$$

Therefore, to whiten a signal X, we calculate its covariance matrix and then decompose it as an eigenvalue such that:

$$R_X = E\{ X(t)X^*(t) \} = E\{ AS(t)(AS(t))^* \}$$
$$= AA^* \underbrace{E\{ S(t)S^*(t) \}}_{I} = AA^* = UDU^T \tag{6.4}$$

where U is the orthogonal matrix of the eigenvectors of matrix R_X and D is the diagonal matrix of the eigenvalues of matrix R_X.

We then obtain:

$$R_X = E\{ WX(t)(WX(t))^* \} = WW^* \underbrace{E\{ X(t)X^*(t) \}}_{UDU^T} = I \tag{6.5}$$

We therefore determine the whitened matrix as follows:

$$W = D^{-\frac{1}{2}}U^T \tag{6.6}$$

After undergoing these preparatory steps, the vector for the observed signals X(t) is used as input for ICA in order to separate the sources via the following steps:

– maximization of the Kurtosis function;

– determination of the optimal step;

– deflation.

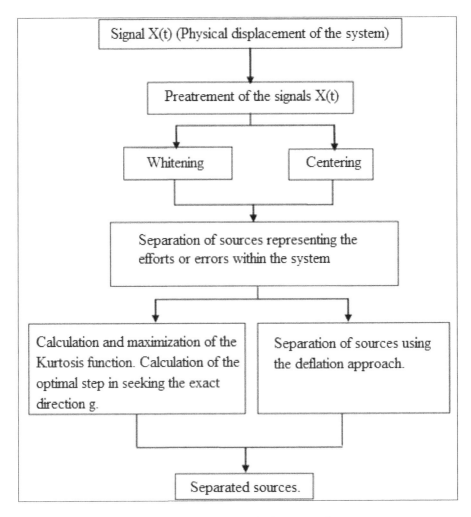

Figure 6.4. *The principal stages in the separation process*

Deflation makes it possible to extract sources one after another. This approach is based on two steps. The first consists of identifying one source from the mix and the second subtracts the source's contribution to the mix. Otherwise, we can use the deflation approach, which consists of calculating the contribution of the source, estimated through observation, using the minimum solution of the squared error average for the linear regression problem.

In our case, the separation program is used to identify the irregularities of the road surface that cause the vibratory response X(t). Numerically, this signal is obtained using the Newmark method. Experimentally, the signal is identified using sensors placed near the outputs of the system, for example, accelerometers.

The principal stages in the separation process are summarized in Figure 6.4.

6.5. Monte Carlo technique

The MC algorithm is a method for solving systems with uncertain parameters. Taking into account the distribution of the probability of parameters, which can be uniform or normal, this method generates samples for each uncertain parameter. It first calculates an estimate of the surface without uncertainty and then calculates the estimate for each iteration with uncertainty. Finally, the error between the estimate with and without uncertainty is evaluated (Dammak et al. 2019). The MC method is applied in this study to evaluate the robustness of ICA with uncertainty in the different parameters, namely, the mass of the driver, the stiffness of the suspension and the variation in the shock absorption, keeping track of the effect of the wind (noise). This is a simple and flexible probabilistic method.

6.6. Results and discussion

In this section, the results of the robustness of the ICA under uncertain conditions are presented.

6.6.1. *Implementation of the ICA approach*

ICA was used in this study to determine the road surface. The steps for applying ICA to the different models studied are presented as follows:

1) Determine the vibratory responses X(t) through numeric analysis using the Newmark approach.

2) These responses represent the observed signals in the ICA.

These stages are presented in Figure 6.5.

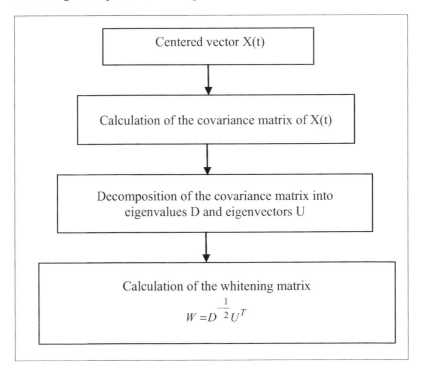

Figure 6.5. *Steps for whitening the vector X(t)*

Then, based on these responses, ICA is used to obtain the separated signals by following the steps presented in the flow chart (see Figure 6.6).

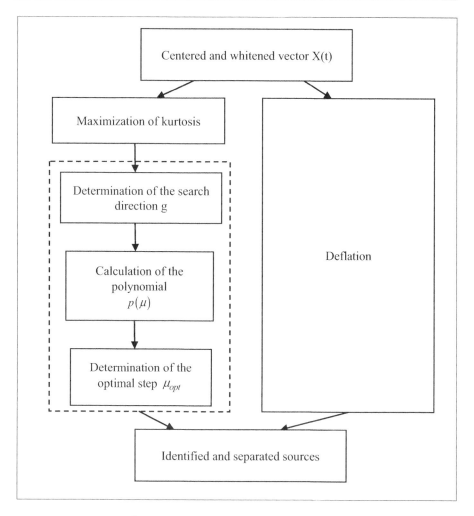

Figure 6.6. *Steps for separating sources*

6.6.2. Influence of noise

For the purpose of showing the influence of the wind, which is introduced as a noise in the quarter-vehicle model, two road profiles were studied, B and D. In this work, the values of the signal to noise ratio (SNR) proposed by

Ben Hassen et al. (2019) were considered. The first, with a noise of 0.9 dB, was added and the uncertain parameters (mass, rigidity and shock absorption) varied by 60% (Table 6.1). The second case, with a noise of 3.5 dB, was considered with the same variation in the uncertain parameters (Table 6.2). Finally, a noise of 9.5 dB was considered (Table 6.3).

Figure 6.7 shows an example of road surface estimates before applying the Monte Carlo technique.

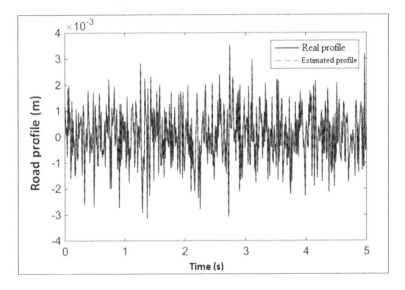

Figure 6.7. *Comparison between the real surface D and the estimated surface. For a color version of this figure, see www.iste.co.uk/elhami/uncertainty.zip*

The results demonstrate that the estimate becomes more sensitive when there is a large amount of noise (9.5 dB), in particular for surface D. Despite this sensitivity, the error values remain acceptable for all the estimated road surfaces, whatever the value of the noise, such that we can conclude that a surface is estimated appropriately using ICA.

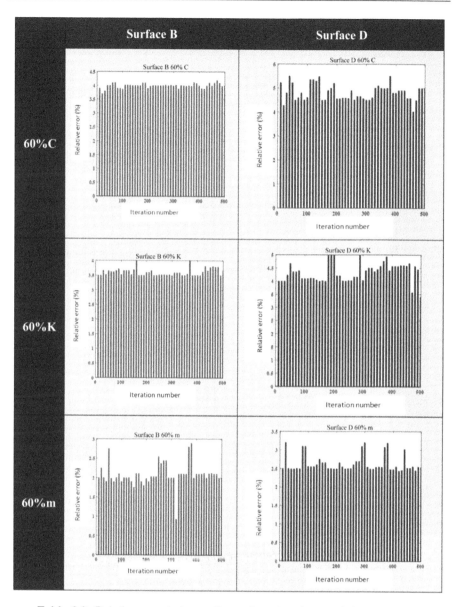

Table 6.3. *Relative error between the real road surface and the estimated surface in the presence of noise at 0.9 dB. For a color version of this table, see http://www.iste.co.uk/elhami/uncertainty.zip*

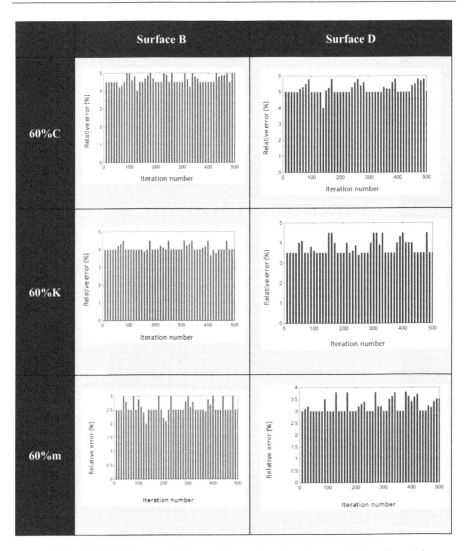

Table 6.4. *Relative error between the real road surface and the estimated surface in the presence of noise at 3.5 dB. For a color version of this table, see http://www.iste.co.uk/elhami/uncertainty.zip*

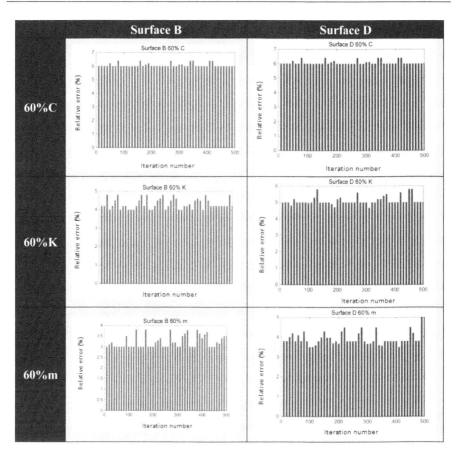

Table 6.5. *Relative error between the real road surface and the estimated surface in the presence of noise at 9.5 dB. For a color version of this table, see http://www.iste.co.uk/elhami/uncertainty.zip*

6.6.3. *Influence of speed*

In this section, a variation of 20 km/h and 100 km/h in the speed of the vehicle was considered to study its influence on the estimate of the road surface. The results of the error between the real surface and the estimated surface are presented in Table 6.6.

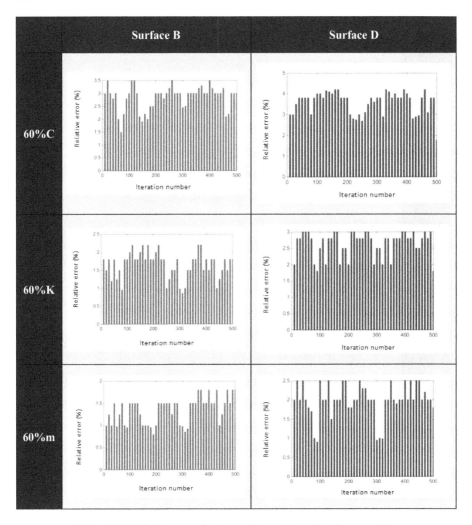

Table 6.6. *Relative error between the real road surface and the estimated surface at a speed of 20 km/h. For a color version of this table, see http://www.iste.co.uk/elhami/uncertainty.zip*

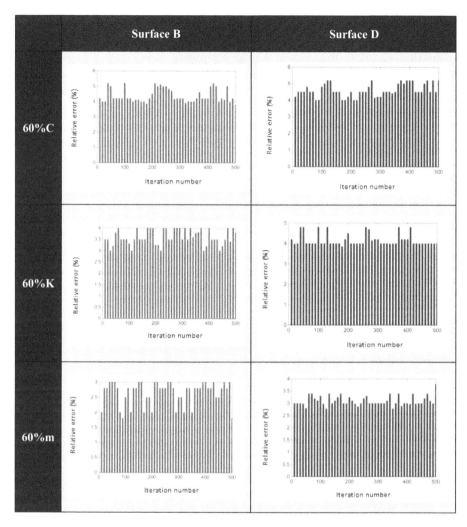

Table 6.7. *Relative error between the real road surface and the estimated surface at a speed of 20 km/h and 100 km/h. For a color version of this table, see http://www.iste.co.uk/elhami/uncertainty.zip*

We see that despite the variation in speed, ICA can estimate a road surface. The maximum relative error did not surpass 6%.

6.7. Conclusion

In this chapter, the Monte Carlo technique was used alongside the Newmark method to solve the movement equation of a quarter-vehicle model. The road surface was then estimated using ICA when the mass of the vehicle, the coefficient of shock absorption and the rigidity of the spring were uncertain, taking the effect of both wind and speed into account. The results obtained depended on the sample used. An increase of this order correlates with a better solution. This confirms the efficiency of ICA in the reconstruction of road disturbances.

6.8. References

Abbes, M.S., Chaabane, M.M., Akrout, A., Fakhfakh, T., Haddar, M. (2011). Vibratory behavior of a double panel system by the operational modal analysis. *International Journal of Modeling, Simulation, and Scientific Computing*, 2(04), 459–479.

Ben Hassen, D., Miladi, M., Abbes, M.S., Baslamisli, S.C., Chaari, F., Haddar, M. (2019). Road profile estimation using the dynamic responses of the full vehicle model. *Applied Acoustics*, 147, 87–99.

Ben Jdidia, A., Ben Hassen, D., Hentati, T., Abbes, M.S., Haddar, M. (2022). Robustness study of the road profile estimation technique under uncertainty. *Journal of Theoretical and Applied Mechanics*, 60(3), 521–533. doi: 10.15632/jtam-pl/151940.

Chaabane, M.M., Ben Hassen, D., Abbes, M.S., Baslamisli, S.C., Chaari, F., Haddar, M. (2019). Road profile identification using estimation techniques: Comparison between independent component analysis and Kalman filter. *Journal of Theoretical and Applied Mechanics*, 57(2), 397–409.

Chitroub, S. (2007). Analyse en composantes indépendantes d'images multibandes : faisabilité et perspectives. *Revue de Télédétection*, 7(1–2), 3–4.

Dammak, K., Koubaa, S., El Hami, A., Walha, L., Haddar, M. (2019). Numerical modeling of uncertainty in acoustic propagation via generalized polynomial chaos. *Journal of Theoretical and Applied Mechanics*, 57(1), 3–15.

Fauriat, W., Mattrand, C., Gayton, N., Beakou, A., Cembrzynski, T. (2016). Estimation of road profile variability from measured vehicle responses. *Vehicle System Dynamics*, 54(5), 585–605.

Haddar, M., Chaari, R., Baslamisli, S.C., Chaari, F., Haddar, M. (2019). Intelligent PD controller design for active suspension system based on robust model-free control strategy. *Proceedings of the Institution of Mechanical Engineers, Part C: Journal of Mechanical Engineering Science*, 233(14), 4863–4880.

Healey, A.J., Nathman, E., Smith, C.C. (1977). An analytical and experimental study of automobile dynamics with random roadway inputs. *Journal of Dynamic Systems, Measurement, and Control*, 99(4), 284–292.

Nodeh, T.F., Mirzaei, M., Khosrowjerdi, M.J. (2021). Simultaneous output selection and observer design for vehicle suspension system with unknown road profile. *IEEE Transactions on Vehicular Technology*, 70(5), 4203–4211.

Taktak, M., Tounsi, D., Akrout, A., Abbès, M.S., Haddar, M. (2012). One stage spur gear transmission crankcase diagnosis using the independent components method. *International Journal of Vehicle Noise and Vibration*, 8(4), 387–400.

Xue, K., Nagayama, T., Zhao, B. (2020). Road profile estimation and half-car model identification through the automated processing of smartphone data. *Mechanical Systems and Signal Processing*, 142, 106–722.

Multi-Objective Optimization Applied to a High Electron Mobility Transistor

In this chapter, we study the high electron mobility transistor (HEMT), which is one of the most important components in high-power mechatronic systems. The HEMT is the most frequently used technology in complex systems in general, and in mechatronic systems in particular. Consequently, the optimization of this technology constitutes a major stake for engineers and researchers in this field. In this chapter, we present a multi-objective optimization method applied to HEMT to improve its thermal and mechanical performance. The optimization process is based on the coupling of two models: the finite element model using Comsol Multiphysics® software and an optimization model written with Matlab® software. The first model is used to simulate the thermomechanic behavior of the HEMT. The second model is used to solve the problem of optimization by coupling it with the first model. The optimal values of the design variables obtained after the application of the optimization process allow us in turn to optimize the thermomechanical behavior of the HEMT structure.

7.1. Introduction

The high electron mobility transistor based on gallium nitride (GaN) has generally been used in high power systems and mechatronic systems. Thanks to its structure and its particular electro-thermal characteristics, compared to

Chapter written by Rabii El Maani, Abdelhamid Amar, Bouchaïb Radi and Abdelkhalak El Hami.

other transistors, HEMT technology can function in exceptional conditions such as high temperatures, high frequency, and high power. Thanks to all of these advantages, the HEMT appears in several fields such as telecommunications, electronic warfare, and satellites (Chen et al. 2019).

When the HEMT is in operation, it dissipates power in the active zone (the gate output next to the drain). This phenomenon provokes an increase in HEMT temperature due to the intrinsic heating of the component (Amar et al. 2022). Although most of the characteristics – such as thermal conductivity and electron mobility – are dependent on temperature, they can be influenced or deteriorated by an increase in temperature within its structure (Amar et al. 2021a; Dundar and Donmezer 2019). This can also generate mechanical degradation in the structure since some physical characteristics of the materials depend simultaneously on the temperature and the deformation. These degradations can take different forms: constraints, displacements or deformations. The identification and understanding of these degradations remain an important challenge in ensuring the reliability and performance of HEMT technology (Amar et al. 2021b).

The task of multi-objective optimization is not to find an optimal solution corresponding to each objective function, but to find a set of solutions called the Pareto-optimal front (Coello 2006; Deb 2014; Konak et al. 2006). Using optimization techniques is crucial for real-world applications due to the multiplicity of implied principles. Most real-world applications depend on several variables in any given model. The appropriate analysis of the variables involved is the key to a successful optimization.

When the objective function of an optimization problem is nonlinear and undifferentiable, evolutionary algorithm (EA) techniques are generally used to find the overall optimum (Deb 2014; El Maani 2020). In an EA, protecting the genetic diversity of a population is very important for the capacity of the population to sustain its iterative development. And generally, the genetic diversity of a population is the result of basic genetic processes such as recombination, crossing, mutation, selection and adaptation (Deb et al. 2002).

In this chapter, we use the BSAMO algorithm, which develops the backtracking search algorithm (BSA), proposed for mono-objective optimization (Civicioglu 2013) when analyzing multi-objective optimization problems with two principal tools: the rapid, non-dominated sorting procedure and the crowding distance. The proposed algorithm and its application to mechatronic systems can be considered a multi-disciplinary optimization. This is a field of engineering that emphasizes the use of numeric optimization in designing systems that cover several disciplines or subsystems.

In order to study HEMT technology using the new approach, this chapter consists of several sections: firstly, we begin with a description of HEMT technology and its uses; then, we discuss multi-physical modeling of the component; in the following section, we define the optimization method used in this study; and the final section is reserved for the presentation and discussion of the results obtained when applying the multi-objective optimization method to the HEMT.

7.2. Description of HEMT technology

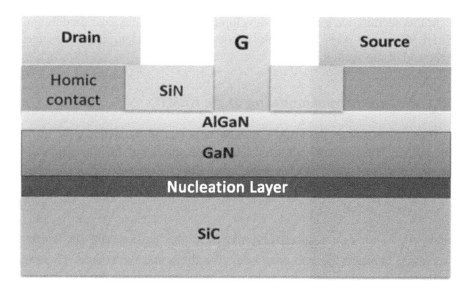

Figure 7.1. *HEMT structure. For a color version of this figure, see www.iste.co.uk/elhami/uncertainty.zip*

HEMT technology is characterized by a particular structure composed of several layers of different materials (Figure 7.1). It begins with a layer made of silicon carbide, which is the substrate upon which the component is built, a nucleation layer that provides mesh alignment between the substrate and the GaN, a layer of GaN that contains a 2DEG gas in the superior portion and a layer of AlGaN that constitutes a heterojunction with the layer of GaN, and other layers that provide functions whose number and materials depend on the manufacturer as well as the applications it will be used for. After a bibliographic study, we noted that a HEMT can have several structures depending on the fields of application. The HEMT is composed of three electrodes: the source, the drain and the gate. They provide control of the current and operational tension of the component. The source and drain contacts are measured in ohms. In this study, we will discuss the single-finger HEMT whose structural dimensions are: $th_{SiC} = 100$ µm, $th_{GaN} = 1.3$ µm and $th_{AlGaN} = 0.03$ µm. These are the widths of the layers of substrate, GaN, and AlGaN, respectively. The width of the gate is $W_{Gate} = 100$ µm.

7.3. Multi-physical modeling of the HEMT

In this section, we look at the development of multi-physical modeling of the HEMT in two parts. The results of the numeric simulations are presented at the end of the section.

7.3.1. *Electro-thermal modeling of the HEMT*

When functioning, the HEMT dissipates power in the active zone at the gate output. This power depends on the tension applied to the boundaries and the current circulating in the transistor, such that:

$$P_{diss} = V_{ds}.I_{ds} \qquad [7.1]$$

where V_{ds} and I_{ds}, respectively, represent the tension and the current circulating between the two electrodes, the drain and the source. To study the thermal behavior of the transistor, we must approach the thermal transfer inside of the component. This transfer happens principally through conduction, whereas the other modes (convection and radiation) are negligible since they only represent less than 1.5% (Aubry 2004). Based on this mode of transfer, the heat equation is expressed as follows (Amar et al. 2021b):

$$k.\nabla^2 T = \rho \cdot C_{P.} \frac{\partial T}{\partial t} + Q \qquad [7.2]$$

where Q is the dissipated power (J); K is the thermal conductivity $(W.m^{-1}.K^{-1})$; ρ is the density $(Kg.m^{-3})$; C_p is the specific heat capacity $(J.Kg^{-1}.K^{-1})$; T is the temperature (K).

This equation relates the dissipated power to the thermal parameters. It also makes it possible to describe the thermal behavior of the transistor in terms of the operational conditions, specifically the dissipated power. The system is composed of several materials whose properties are presented in Table 7.1, with some of the properties depending on the temperature (Marcon et al. 2013).

Materials	Volumetric mass (density), ρ (kg/m³)	Thermal conductivity, K (W/m/K)	Specific heat capacity, Cp (J/kg/K)
Au	19,300	310	137
SiN	3,300	10	713
AlGaN	5,470	$25.\left(\dfrac{293}{273+T}\right)^{-1.35}$	548
GaN	6,100	$161.\left(\dfrac{293}{273+T}\right)^{-1.45}$	490
SiC	3,220	$416.\left(\dfrac{293}{273+T}\right)^{-1.5}$	690

Table 7.1. *Thermal properties of materials*

A finite element model was developed using the Comsol Multiphysics software, taking the previous properties into account. After making the electro-thermal model, the results of numerical simulations were obtained; these are presented in Figure 7.2. Moreover, we present a comparison with experimental results. From this figure, we noted that the operational temperature of the HEMT rises with the increase in dissipated power, following a nonlinear evolution. We can also see a good correlation between the experimental results and those of the simulations. This criterion allows us to validate our model.

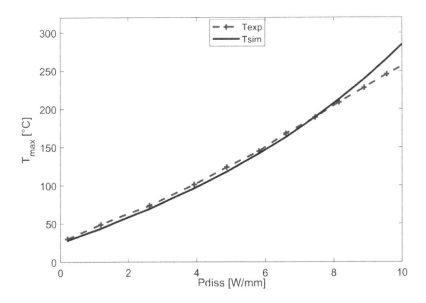

Figure 7.2. *Evolution in temperature relative to dissipated power. For a color version of this figure, see www.iste.co.uk/elhami/uncertainty.zip*

7.3.2. *Thermomechanical modeling of the HEMT*

In this section, we examine thermomechanical modeling, which allows us to study the impact of thermal behavior on the mechanical behavior of HEMT technology. The increase in temperature in the HEMT due to the dissipated power can lead to a mechanical degradation of its structure. To identify these degradations, we develop a thermomechanical model that ties the thermal parameters to the mechanical parameters. The temperature is the most important thermal parameter. The mechanical parameters can take many forms including displacement, deformations and constraints.

The thermal model can be coupled with the mechanical model using thermo-elastic equations. Deformations and entropy density can be defined using the following expressions (Amar et al. 2021b):

$$\{\varepsilon\} = [D]^{-1}\{\sigma\} + \{\alpha\}\Delta T \qquad [7.3]$$

$$\{S\} = \{\alpha\}^{T}\{\sigma\} + \frac{\rho \cdot C_{P}}{T_{0}}\Delta T; \Delta T = T - T_{\text{ref}} \qquad [7.4]$$

where $\{\varepsilon\}$ is the vector for total deformation. It is dependent on the vector for constraints, the vector for the thermal expansion coefficients, the difference in temperature and the inverse of the elastic rigidity matrix.

In equation [7.4], the entropy depends on several thermal and mechanical parameters such as the constraints and variation in temperature, among others.

According to the laws of thermodynamics, the density of entropy can be expressed as a function of the thermal flux, following this equation:

$$Q = T_0 \cdot S.$$ [7.5]

Meanwhile, the vector for thermo-elastic coefficients is expressed as follows:

$$\{\beta\} = [D]\{\alpha\}$$ [7.6]

Denoting the specific heat capacity at a constant volume C_v, the expression is:

$$C_v = C_p - \frac{T_0}{T}\{\alpha\}^T\{\beta\}$$ [7.7]

Integration of the previous equations allows us to obtain new forms of the principal equations, which therefore become (Amar et al. 2021b):

$$\{\varepsilon\} = [D]\{\varepsilon\} - \{\beta\}\Delta T$$ [7.8]

$$Q = T_0\{\beta\}^T\{\varepsilon\} + \rho \cdot C_v \cdot \Delta T$$ [7.9]

Materials	E (GPa)	CTE (1/K)	U
Au	70	14.2	0.44
SiN	250	$1.5e^{-6}$	0.27
AlGaN	212.75	$5.2425e^{-6}$	0.319
GaN	181	$8.6e^{-6}$	0.352
SiC	748	$4.3e^{-6}$	0.45

Table 7.2. *Physical properties of materials*

These two equations allow for thermomechanical coupling: they allow us to relate the thermal parameters to the mechanical ones. The two elementary couplings, electro-thermal and thermomechanic, create the overall electro-thermomechanical coupling. The physical properties of the necessary materials for realizing the model are listed in Table 7.2 (El Hami and Pougnet 2015). Some properties are already integrated in the library of materials that comes with Comsol Multiphysics.

The finite element model that we developed with the Comsol Multiphysics software allowed us to observe the mechanical behavior in general and the distribution of the von Mises constraints within the HEMT structure. This distribution is presented in Figure 7.3: we can see that the constraints are high within the substrate, which is the base for the construction of the component, and on both sides of the gate.

These constraints increase with the dissipated power, following a nonlinear evolution, as shown in Figure 7.4. They reach a value of 4.4 GPa in the critical zones of the structure. Consequently, the operational conditions of the component have a great mechanical impact on the structure of the HEMT.

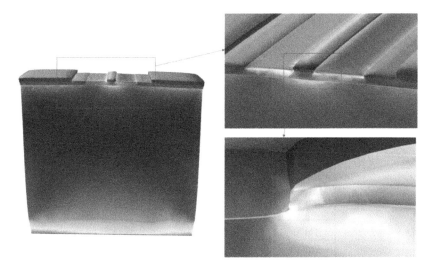

Figure 7.3. *Distribution of von Mises constraints in the HEMT structure. For a color version of this figure, see www.iste.co.uk/elhami/uncertainty.zip*

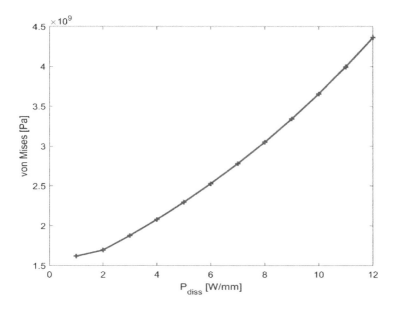

Figure 7.4. *Evolution of von Mises constraints relative to dissipated power. For a color version of this figure, see www.iste.co.uk/elhami/uncertainty.zip*

7.4. The multi-objective optimization approach

Optimization is a process that makes it possible to improve an already existent system or to determine the parameters that can lead to the design of a new system with a minimal cost and high quality. The problem of multi-objective optimization can be expressed as follows:

$$
\begin{cases}
\displaystyle\min_{x\in\Omega} \mathbf{f}(x) = (f_1(x),\ldots,f_m(x))^T \\
\text{s.c}: \quad g_i(x) \le 0, \quad \text{for} \quad i = 1,\ldots, l \\
\qquad\quad h_k(x) = 0, \quad \text{for} \quad k = 1,\ldots, p
\end{cases}
\qquad [7.10]
$$

where $\mathbf{f}$ is the vector for the concurrent objective functions that are to be optimized, m is the number of objective functions, $x = (x_1,\cdots,x_n) \in \Omega$ is the decision space in D-dimensions where each decision variable x_1 is bound by inferior and superior limits $x_{li} \le x_i \le x_{ui}$ for $i = 1,\ldots,D$, g_i (x) are l inequality constraints, and h_k (x) are p equality constraints.

The BSA was initially only developed for optimization problems with a unique objective. In this chapter, we wish to apply an appropriate extension of it so it can deal with multi-objective optimization problems and solve a mechatronic problem in a high electron mobility transistor. To do this, the principal development should address the values of physical conditions attributed to individuals in the population.

Algorithm 7.1. Pseudocode of the BSAMO

Function $P = BSAMO(D, N, m, \mathbf{f}(x), \text{Max}_G, x_l, x_u)$

1: Generate the initial population $\mathbf{P}$ using 1 and the historical population $\mathbf{P}_h$

2: $\mathbf{P} \coloneqq$ Find_Non_Dominated_Crowding_Distance $(\mathbf{P} \mid \mathbf{f}(x))$

3: **for** $t = 1 : \text{Max}_G$, **do**

4: $[\mathbf{P}_c, \mathbf{P}_h] \coloneqq$ BSA_Operator$(\mathbf{P}, x_l, x_u, \mathbf{P}_h)$; // BSA_Operator : Mutation defi and Crossover in 6

5: $\mathbf{P} \coloneqq$ Find_Non_Dominated_Crowding_Distance $([\mathbf{P}; \mathbf{P}_c] \mid \mathbf{f}(x))$

6: Sort and find the current Pareto optimal solution

7: **end for**

Indeed, this development adapts the principal strategies and the simplicity of the BSA algorithms to get the most out of new ways of navigating through the sphere of research and using directional matrices for adapted research, providing good efficiency and preserving the spirit of the BSA.

Algorithm 7.1 gives the proposed code nickname, where Max_G is the maximal number of generations. It is called the BSAMO (El Maani et al. 2019). It integrates the mutation and crossing operators from the BSA and the fast, non-dominated sorting and the crowding distance from Deb et al. (2002).

The following sections provide a few explanations regarding this last element.

7.4.1. *Fast non-dominated sorting*

The procedure of fast, non-dominated sorting was developed in the context of the NSGA-II (Deb et al. 2002). While creating this procedure, the domination count np, the number of solutions dominating solution p, and the entire set of solutions Sp that solution p dominates are calculated for each solution. The first, non-dominated, front is thus created and initialized with all the solutions whose domination count is zero. Then, for each solution p with np = 0, each member q of the set Sp is revisited, and its domination number is reduced by one. Consequently, if for any member, the domination count is equal to zero, it is then placed in a separate list Q. The second non-dominated front is then created as a union of all the individuals belonging to Q. The procedure is repeated for the next fronts (F3, F4, etc.) until all the individuals have been assigned to their ranks. Fitness is defined by level numbers, with lower numbers corresponding to a higher fitness (F1 is the best).

7.4.2. Crowding distance

The crowding distance defined in Deb et al. (2002) is used as an estimate of the measure of diversity among individuals surrounding a given individual i in the population. This distance is the average distance between two individuals situated on either side of the given solution along the length of each objective. Figure 7.1 illustrates the average distance between individuals $i-1$ and $i+1$ surrounding individual i situated at the Pareto front. This distance is an estimate of the perimeter of the cuboid formed using the closest neighbors. This metric represents half of the perimeter of the cuboid containing solution i.

The principal consideration in the crowding distance is to find the Euclidian distance between each individual on a front as a function of their m objectives. The crowding distance calculation, based on the normalized objective values, is given by Algorithm 7.2, where f_m^{max} and f_m^{min} are, respectively, the maximal and minimal values of the m-th objective function. The sum of each crowding distance value corresponding to each objective gives the overall value of the crowding distance.

In this algorithm, $\mathcal{I}$ is a non-dominated set, n is the number of elements in $\mathcal{I}$, $\mathcal{I}[i]_m$ is the m-th objective value of the individual in $\mathcal{I}$, and the sorting $(\mathcal{I}, m)$ is the sorting of individuals $\mathcal{I}$ following to the m-th objective. The theoretical aspect of this algorithm is discussed in El Maani (2020).

Algorithm 7.2. Crowding distance calculation for a set solution
$1: n = \ \|\mathcal{I}\| \ // \text{ number of solutions in } \mathcal{I}$
$2: \textbf{for } \text{each } i, \ \textbf{do}$
$3: \text{set } \mathcal{I}[i]_{\text{distance}} = 0$
$4: \textbf{end for}$
$5: \textbf{for } \text{each objective } m, \ \mathcal{I} = \text{sort}(\mathcal{I}, m) \ \textbf{do}$
$6: \mathcal{I}[1]_{\text{distance}} = \mathcal{I}[n]_{\text{distance}} = \infty$
$7: \quad \textbf{for } i = 2 \text{ to } (n-1) \ \textbf{do}$
$8: \mathcal{I}[i]_{\text{distance}} = \mathcal{I}[i]_{\text{distance}} + (\mathcal{I}[i+1]_{.m} - \mathcal{I}[i-1]_{.m}) / (f_m^{max} - f_m^{min})$
$9: \quad \textbf{end for}$
$10: \textbf{end for}$

7.5. Multi-objective optimization applied to HEMT technology

7.5.1. *Description of the optimization problem*

When the HEMT is in operation, the interaction between electric and thermal phenomena can give rise to thermal as well as mechanical phenomena within its structure. These can take the form of displacement, deformation and constraints. These phenomena can influence electron performance and lead to transistor failure. Consequently, control and minimization of the effect of these phenomena represents a stake that greatly affects the reliability of the component. In this sense, the goal of this study is to minimize the effect of the operational temperature and that of the von Mises constraints over the HEMT structure under determinist constraints. The optimization problem is addressed using the multi-objective optimization approach. Consequently, the optimization problem is expressed as follows:

$$\begin{cases} \min_{x} f(x) = \{\sigma_e(x), T(x)\} \\ \qquad\quad 90 \leq a \leq 160 \\ s.c: \quad 1.1 \leq b \leq 1.9 \\ \qquad\quad 0.01 \leq c \leq 0.09 \end{cases} \qquad [7.11]$$

where x is the vector of determinist design variables, $\sigma_e(x)$ are the von Mises constraints of the structure, and $T(x)$ is the operational temperature.

The execution of this process requires a coupling of two models: the finite element model, which we built using Comsol Multiphysics software, and the multi-objective optimization model, which we developed using Matlab software. We chose three geometric parameters – the thickness of the substrate (a), the thickness of the GaN (b) and the thickness of the AlGaN (c) – as design variables since they have a great influence on the electro-thermomechanical behavior of the HEMT. Then, we ranked the initial, minimal and maximal values needed to solve this problem, as shown in Table 7.3.

Parameters (µm)	Deterministic	Minimal	Maximal
a	100	90	160
b	1.7	1.1	1.9
c	0.03	0.01	0.09

Table 7.3. *Results of the HEMT reliability analysis*

7.5.2. *Results and discussion*

Figure 7.5 and Table 7.4 show, respectively, the Pareto solutions obtained using the BSAMO algorithm and the best optimal functions of the optimal design values of the HEMT, also solved for an optimization process combined with a population of 30 individuals.

The results obtained through BSAMO are presented in Table 7.4. This table shows the optimal values of the design variables for the minimal temperature value and those for a minimal von Mises constraint value within the HEMT structure. The Pareto front indicates the other individuals that also present a good, minimizing compromise between the objective functions with other, very specific design variables.

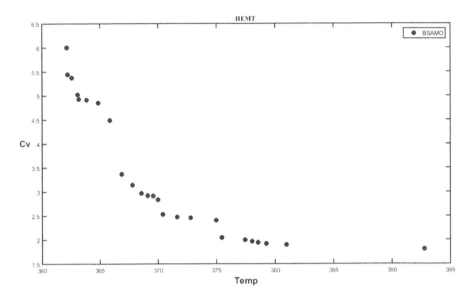

Figure 7.5. *Pareto solutions to the HEMT problem. For a color
version of this figure, see www.iste.co.uk/elhami/uncertainty.zip*

	Design variables			Objective functions	
	a	b	c	Temperature	von Mises constraint
BSAMO	100	1.86528350	0.01	**362.13466164**	6.00046839
	127.39324702	1.90	0.09	392.77624616	**1.80605790**

Table 7.4. *Optimal results for the transistor*

7.6. Conclusion

In this chapter, we proposed an approach for improving the
thermomechanical performance of the HEMT using a multi-objective
optimization approach. The application of this approach is possible thanks to
the coupling of two models: the first is the finite element model, developed
using Comsol Multiphysics software, and the second is the multi-objective
optimization model, written using Matlab software. Its purpose is to
minimize two objective functions, temperature and the von Mises constraint,
using precisely determined design variables. The application of this
approach to HEMT technology allowed us to explore new design parameters

that satisfy a Pareto front that provides a good compromise of minimal temperature and von Mises constraint values.

7.7. References

Amar, A., Radi, B., El Hami, A. (2021a). Electrothermal reliability of the high electron mobility transistor (HEMT). *Applied Sciences (Switzerland)*, 11(10720). doi: 10.3390/app112210720.

Amar, A., Radi, B., El Hami, A. (2021b). Reliability based design optimization applied to the high electron mobility transistor (HEMT). *Microelectronics Reliability*, 124(July), 114299.

Amar, A., Radi, B., El Hami, A. (2022). Optimization based on electro-thermomechanical modeling of the high electron mobility transistor (HEMT). *International Journal for Simulation and Multidisciplinary Design Optimization*, 13(2). doi: 10.1051/smdo/2021035.

Aubry, R. (2004). Étude des aspects électrothermiques de la fikière HEMT AlGaN/GaN pour application de puissance hyperfréquence. PhD Thesis, Université des sciences et technologiques de Lille, Lille.

Chen, Y., Xu, Y., Wang, F., Wang, C., Zhang, Y., Yan, B., Xu, R. (2019). Improved quasi-physical zone division model with analytical electrothermal I_{ds} model for AlGaN/GaN heterojunction high electron mobility transistors. *International Journal of Numerical Modelling: Electronic Networks, Devices and Fields*, 33(3), 1–17. doi: 10.1002/jnm.2630.

Civicioglu, P. (2013). Backtracking search optimization algorithm for numerical optimization problems. *Applied Mathematics and Computation*, 15, 8121–8144.

Coello, C. (2006). Evolutionary multi-objective optimization: A historical view of the field. *IEEE Computational Intelligence Magazine*, 1, 28–36.

Deb, K. (2014). Multi-objective optimization. In *Search Methodologies*, Burke, E. and Kendall, G. (eds). Springer, Boston.

Deb, K., Pratap, A., Agarwal, S., Meyarivan, T. (2002). A fast and elitist multiobjective genetic algorithm: NSGA-II. *IEEE Transactions on Evolutionary Computation*, 2, 182–197.

Dundar, C. and Donmezer, N. (2019). Thermal characterization of field plated AlGaN/GaN HEMTs. In *InterSociety Conference on Thermal and Thermomechanical Phenomena in Electronic Systems, ITHERM*, May 2019, 755–760. doi: 10.1109/ITHERM.2019.8757323.

El Hami, A. and Pougnet, P. (2015). *Embedded Mechatronic Systems 2: Analysis of Failures, Modeling, Simulation and Optimization*. ISTE Press, London, and Elsevier, Oxford.

El Maani, R., Radi, B., El Hami, A. (2019). Multiobjective backtracking search algorithm: Application to FSI. *Structural and Multidisciplinary Optimization*, 59, 131–151.

El Maani, R., Radi, B., El Hami, A. (2020). Multiobjective aerodynamic shape optimization of NACA0012 airfoil based mesh morphing. *International Journal for Simulation and Multidisciplinary Design Optimization*, 11, 10.

Konak, A., Coit, D.W., Smith, A.E. (2006). Multi-objective optimization using genetic algorithms: A tutorial. *Reliability Engineering & System Safety*, 9, 992–1007.

Lorin, E. and Stevens (2013). Thermo-piezo-electro-mechanical simulation of AlGaN (aluminum gallium nitride)/GaN (gallium nitride) high electron mobility transistor. Graduate thesis, Utah State University, Logan.

Marcon, D., Meneghesso, G., Wu, T., Stoffels, S., Meneghini, M., Zanoni, E., Decoutere, S. (2013). Reliability analysis of permanent degradations on AlGaN/GaN HEMTs. *IEEE Transactions on Electron Devices*, 60(10), 3132–3141.

List of Authors

Mohamed Slim ABBES
National Engineering School of Sfax
(ENIS)
Tunisia

Abdelhamid AMAR
IRDL CNRS UMR 6027
University of Southern Brittany
Lorient
France

Hicham BAAMMI
Northvolt
Skellefteå
Sweden

Dorra BEN HASSEN
National Engineering School of Sfax
(ENIS)
Tunisia

Anoire BEN JDIDIA
National Engineering School of Sfax
(ENIS)
Tunisia

Fakher CHAARI
National Engineering School of Sfax
(ENIS)
Tunisia

Abel CHEROUAT
University of Technology of Troyes
(UTT)
France

Abdelkhalak EL HAMI
INSA Rouen Normandie
Saint-Étienne-du-Rouvray
France

Rabii EL MAANI
ENSAM
Meknes
Morocco

Maroua HADDAR
National School of Engineers
of Sousse (ENISo)
Tunisia

Mohamed HADDAR
National Engineering School of Sfax
(ENIS)
Tunisia

Ghais KHARMANDA
3D Printing 4U (UG)
Cologne
Germany

Bouchaïb RADI
Faculté des sciences et techniques
(FST)
Université Hassan I
Settat
Morocco

Mohamed TAOUFIK KHBOU
National Engineering School of Sfax
(ENIS)
Tunisia

Index

3D printing, 24

A, B

accelerated aging, 42
acoustic emission (AE), 39, 43–46,
    52, 70, 72, 73, 75
    continuous, 45
    discrete, 44
additive manufacturing, 19–21,
    24–27, 29, 30, 32, 34, 35
agro-composite, 41, 42, 62, 63, 65,
    67, 68, 74, 75, 86–89
algebraic estimator, 103, 109, 114,
    115, 138
artificial intelligence (AI), 19–24, 27,
    29, 32, 35
bio-sourced materials, 40, 41

C, D

centering, 6, 142, 143
composite, 39–45, 47, 54, 62–65,
    67–74, 76, 86–90
control (*see also* intelligent
    control(ler)), 97–103, 105, 108,
    109, 112–114
    nondestructive, 43, 45, 48
    passive, 100
    semi-active, 99, 100

suspension, 100
    vertical vibration, 108, 109
crowding distance, 159, 166–168
damage, 39, 42, 43, 46–48, 54,
    62–72, 74–76, 86–90
deflation, 9, 144, 145
design, 20, 27
    acceleration, 27
    optimization, 27
diagnostic, 28, 29
durability, 25, 39, 43, 45
dynamic, 2, 3

E, F

eco-production, 1
electric, 119, 121
energy, 1, 2
failure
    cause of, 120, 131
    scenarios, 26, 28, 30, 31, 33–35
feedback, 97, 99, 100, 104
fibers
    carbon, 41
    glass, 41
    hemp, 62, 64, 86, 87, 90
    plant, 40–42, 90
    rupture, 64, 65, 67, 74–76, 86, 87,
        90

Other titles from

in

Mechanical Engineering and Solid Mechanics

2023

EL HAMI Abdelkhalak, DELAUX David, GRZESKOWIAK Henri
*Applied Reliability for Industry 1: Predictive Reliability for the Automobile,
Aeronautics, Defense, Medical, Marine and Space Industries
(Reliability of Multiphysical Systems Set – Volume 16)
Applied Reliability for Industry 2: Experimental Reliability for the
Automobile, Aeronautics, Defense, Medical, Marine and Space Industries
(Reliability of Multiphysical Systems Set – Volume 17)
Applied Reliability for Industry 3: Operational Reliability for the
Automobile, Aeronautics, Defense, Medical, Marine and Space Industries
(Reliability of Multiphysical Systems Set – Volume 18)*

2022

BAYLE Franck
*Product Maturity 1: Theoretical Principles and Industrial Applications
(Reliability of Multiphysical Systems Set – Volume 12)
Product Maturity 2: Principles and Illustrations
(Reliability of Multiphysical Systems Set – Volume 13)*

EL HAMI Abdelkhalak, DELAUX David, GRZESKOWIAK Henri
Reliability and Physics-of-Healthy in Mechatronics
(Reliability of Multiphysical Systems Set – Volume 15)

LANNOY André
Reliability of Nuclear Power Plants: Methods, Data and Applications
(Reliability of Multiphysical Systems Set – Volume 14)

LEDOUX Michel, EL HAMI Abdelkhalak
Heat Transfer 3: Convection, Fundamentals and Monophasic Flows
(Mathematical and Mechanical Engineering Set – Volume 11)
Heat Transfer 4: Convection, Two-Phase Flows and Special Problems
(Mathematical and Mechanical Engineering Set – Volume 12)

PLANCHETTE Guy
Cindynics, The Science of Danger: A Wake-up Call
(Reliability of Multiphysical Systems Set – Volume 11)

2021

CHALLAMEL Noël, KAPLUNOV Julius, TAKEWAKI Izuru
Modern Trends in Structural and Solid Mechanics 1: Static and Stability
Modern Trends in Structural and Solid Mechanics 2: Vibrations
Modern Trends in Structural and Solid Mechanics 3: Non-deterministic Mechanics

DAHOO Pierre Richard, POUGNET Philippe, EL HAMI Abdelkhalak
Applications and Metrology at Nanometer Scale 1: Smart Materials, Electromagnetic Waves and Uncertainties
(Reliability of Multiphysical Systems Set – Volume 9)
Applications and Metrology at Nanometer Scale 2: Measurement Systems, Quantum Engineering and RBDO Method
(Reliability of Multiphysical Systems Set – Volume 10)

LEDOUX Michel, EL HAMI Abdelkhalak
Heat Transfer 1: Conduction
(Mathematical and Mechanical Engineering Set – Volume 9)
Heat Transfer 2: Radiative Transfer
(Mathematical and Mechanical Engineering Set – Volume 10)

2020

SALENÇON Jean
Elastoplastic Modeling

2019

BAYLE Franck
Reliability of Maintained Systems Subjected to Wear Failure Mechanisms:
Theory and Applications
(Reliability of Multiphysical Systems Set – Volume 8)

BEN KAHLA Rabeb, BARKAOUI Abdelwahed, MERZOUKI Tarek
Finite Element Method and Medical Imaging Techniques in Bone
Biomechanics
(Mathematical and Mechanical Engineering Set – Volume 8)

IONESCU Ioan R., QUEYREAU Sylvain, PICU Catalin R., SALMAN Oguz Umut
Mechanics and Physics of Solids at Micro- and Nano-Scales

LE VAN Anh, BOUZIDI Rabah
Lagrangian Mechanics: An Advanced Analytical Approach

MICHELITSCH Thomas, PÉREZ RIASCOS Alejandro, COLLET Bernard,
NOWAKOWSKI Andrzej, NICOLLEAU Franck
Fractional Dynamics on Networks and Lattices

SALENÇON Jean
Viscoelastic Modeling for Structural Analysis

VÉNIZÉLOS Georges, EL HAMI Abdelkhalak
Movement Equations 5: Dynamics of a Set of Solids
(Non-deformable Solid Mechanics Set – Volume 5)

2018

BOREL Michel, VÉNIZÉLOS Georges
Movement Equations 4: Equilibriums and Small Movements
(Non-deformable Solid Mechanics Set – Volume 4)

FROSSARD Etienne
Granular Geomaterials Dissipative Mechanics: Theory and Applications in
Civil Engineering

RADI Bouchaib, EL HAMI Abdelkhalak
Advanced Numerical Methods with Matlab® 1: Function Approximation
and System Resolution
(Mathematical and Mechanical Engineering Set – Volume 6)
Advanced Numerical Methods with Matlab® 2: Resolution of Nonlinear,
Differential and Partial Differential Equations
(Mathematical and Mechanical Engineering Set – Volume 7)

SALENÇON Jean
Virtual Work Approach to Mechanical Modeling

2017

BOREL Michel, VÉNIZÉLOS Georges
Movement Equations 2: Mathematical and Methodological Supplements
(Non-deformable Solid Mechanics Set – Volume 2)
Movement Equations 3: Dynamics and Fundamental Principle
(Non-deformable Solid Mechanics Set – Volume 3)

BOUVET Christophe
Mechanics of Aeronautical Solids, Materials and Structures
Mechanics of Aeronautical Composite Materials

BRANCHERIE Delphine, FEISSEL Pierre, BOUVIER Salima,
IBRAHIMBEGOVIC Adnan
From Microstructure Investigations to Multiscale Modeling:
Bridging the Gap

2014

ATANACKOVIC M. Teodor, PILIPOVIC Stevan, STANKOVIC Bogoljub, ZORICA Dusan
Fractional Calculus with Applications in Mechanics: Vibrations and Diffusion Processes
Fractional Calculus with Applications in Mechanics: Wave Propagation, Impact and Variational Principles

CIBLAC Thierry, MOREL Jean-Claude
Sustainable Masonry: Stability and Behavior of Structures

ILANKO Sinniah, MONTERRUBIO Luis E., MOCHIDA Yusuke
The Rayleigh−Ritz Method for Structural Analysis

LALANNE Christian
Mechanical Vibration and Shock Analysis – 5-volume series – 3rd edition
Sinusoidal Vibration – Volume 1
Mechanical Shock – Volume 2
Random Vibration – Volume 3
Fatigue Damage – Volume 4
Specification Development – Volume 5

LEMAIRE Maurice
Uncertainty and Mechanics

2013

ADHIKARI Sondipon
Structural Dynamic Analysis with Generalized Damping Models: Analysis

ADHIKARI Sondipon
Structural Dynamic Analysis with Generalized Damping Models: Identification

BAILLY Patrice
Materials and Structures under Shock and Impact

BASTIEN Jérôme, BERNARDIN Frédéric, LAMARQUE Claude-Henri
Non-smooth Deterministic or Stochastic Discrete Dynamical Systems:
Applications to Models with Friction or Impact

EL HAMI Abdelkhalak, RADI Bouchaib
Uncertainty and Optimization in Structural Mechanics

KIRILLOV Oleg N., PELINOVSKY Dmitry E.
Nonlinear Physical Systems: Spectral Analysis, Stability and Bifurcations

LUONGO Angelo, ZULLI Daniele
Mathematical Models of Beams and Cables

SALENÇON Jean
Yield Design

2012

DAVIM J. Paulo
Mechanical Engineering Education

DUPEUX Michel, BRACCINI Muriel
Mechanics of Solid Interfaces

ELISHAKOFF Isaac *et al.*
Carbon Nanotubes and Nanosensors: Vibration, Buckling
and Ballistic Impact

GRÉDIAC Michel, HILD François
Full-Field Measurements and Identification in Solid Mechanics

GROUS Ammar
Fracture Mechanics – 3-volume series
Analysis of Reliability and Quality Control – Volume 1
Applied Reliability – Volume 2
Applied Quality Control – Volume 3

RECHO Naman
Fracture Mechanics and Crack Growth

2011

KRYSINSKI Tomasz, MALBURET François
Mechanical Instability

SOUSTELLE Michel
An Introduction to Chemical Kinetics

2010

BREITKOPF Piotr, FILOMENO COELHO Rajan
Multidisciplinary Design Optimization in Computational Mechanics

DAVIM J. Paulo
Biotribology

PAULTRE Patrick
Dynamics of Structures

SOUSTELLE Michel
Handbook of Heterogenous Kinetics

2009

BERLIOZ Alain, TROMPETTE Philippe
Solid Mechanics using the Finite Element Method

LEMAIRE Maurice
Structural Reliability

2007

GIRARD Alain, ROY Nicolas
Structural Dynamics in Industry

GUINEBRETIÈRE René
X-ray Diffraction by Polycrystalline Materials

KRYSINSKI Tomasz, MALBURET François
Mechanical Vibrations

Printed and bound by CPI Group (UK) Ltd, Croydon, CR0 4YY

27/10/2024

14580731-0002